THE BEAVERS OF
POPPLE'S POND

SKETCHES FROM THE LIFE
OF AN HONORARY RODENT

To Helen
and the beavers
of Scotland!

from Patti
and the beavers
of Popples' Pond

Ducky & Growler's
second winter home

Hideaway Pond
6/12 - 5/13
Ducky & Growler
birthplace of Fern

Surprise Pond
10/08 - 7/09
Willow, Bunchberry, Ducky
birthplace of Snowberry

Lake Dismal
8/09 - 9/10
Willow, Bunchberry
Ducky, Snowberry
birthplace of Dewberry

Popple's Pond
4/08 - 9/08
Willow, Popple
Bunchberry
birthplace of Ducky
W, B, S, S & B return 5/13

Beaver camp
Terrible Jack's meadow

Jenny Lake
Ducky's first home
Summer 2010

The Einsteins
Winter 2012/13

The Einsteins
summer of 2011

Ducky & Growler's
first winter

BEAVERS OF MY VALLEY, 2008–2012

NOT TO SCALE

The Balsams
6/11 – 7/12
Willow, Bunchberry
Ducky, Snowberry, Sundew
birthplace of Balsam

The Spoils
10/10 – 5/11
Willow, Bunchberry
Ducky, Dewberry,
Snowberry
birthplace of Sundew

To the Alder Swamp
Mike ➔

Ducky & Growler's
first homestead
summer of 2010

**Einstein
Encampment**
4/10 – 5/11
The Einsteins

THE BEAVERS
OF POPPLE'S
POND

≈

SKETCHES FROM THE LIFE
OF AN HONORARY RODENT

Patti A. Smith

GREEN WRITERS PRESS *Brattleboro, Vermont*

Printed in the United States

10 9 8 7 6 5 4 3 2 1

Giving voice to writers & artists who will make the world a better place

Green Writers Press | Brattleboro, Vermont

www.greenwriterspress.com

ISBN: 978-0-9893104-4-4

Green Writers Press gratefully acknowledges support from individual donors, friends, and readers to help support the environment and our publishing initiative.

COVER ART & BOOK DESIGN

Patti Smith & Dede Cummings

PRINTED IN VERMONT ON PAPER WITH PULP THAT COMES FROM FSC-CERTIFIED FORESTS, MANAGED FORESTS THAT GUARANTEE RESPONSIBLE ENVIRONMENTAL, SOCIAL, AND ECONOMIC PRACTICES. MADE WITH A CHLORINE-FREEPROCESS (ECF: ELEMENTAL CHLORINE FREE).

Contents ≈

∿ PART II: OTHER SKETCHES ∿

We need another and a wider and perhaps a more mystical concept of animals. Remote from universal nature, and living by complicated artifice, man in civilization surveys the creature through the glass of his knowledge and sees thereby a feather magnified and whole image in distortion. We patronize them for their incompleteness, for their tragic fate of having taken a form so far below ourselves. And therein we err and greatly err. For the animal shall not be measured by man. In a world older and more complete than ours they move finished and complete, gifted with extensions of the senses we have lost or never attained, living by voices we shall never hear. They are not brethren, they are not underlings; they are other nations, caught with ourselves in the net of life and time, fellow prisoners of the splendor and travail of earth.

— HENRY BESTON, *The Outermost House*

Introduction

~

WHEN I WAS SIX, I married a cat—a long-haired tomcat the color of smoke. I fell in love with him the first day he showed up at our house, sniffing around for my pet she-cat. For no particular reason, I was sure my feelings were reciprocated. I do not know how many hours I spent following that cat, sometimes on hands and knees, before he snapped—whirling, hissing, and slashing my forehead. When I couldn't get the bleeding to stop, I told my mother I had fallen in a brier bush. She took me to the hospital where they stitched me up. I'm pretty sure the swatting ended that marriage, but it did nothing to dull my fascination with animals of all kinds.

It may be that some of my early literary influences can be blamed for my childhood eccentricities. Beatrix Potter, Thornton Burgess, and E.B. White populated my world with talking trees and friendly fauna. Those were the days when children were ordered to "go out and play." What could be better? Outside were woods and rocks and dirt, the perfect raw materials for a child's imagination.

~

Because one is driven to learn about and understand the beloved, I was drawn to the environmental sciences in college and became a professional naturalist. I moved from tales of squirrels sipping tea before the hearth to a bigger story, one I think of as the "master story," the story from which all others arise—the story of how this ball of rock came to be peopled so extravagantly, a story in which each ephemeral character, each flitting life, has participated in a dance so intricate and densely intertwined that together they have created something solid and enduring. It is a story so ancient that it is impossible to recover all of the pieces, a story so improbable as to seem miraculous. And yet, here we are on this sunlit globe, on this day, you and I and the whole wondrous cargo.

While the trees no longer talk to me, I find that animals still do—in that universal language of posture, expression, and movement. Some speak considerably more clearly than others. On the best occasions, I feel a mutual curiosity. When I sift through my memories for the moments when I have felt deeply honored, there are few that match those that come from being welcomed into the trust of a wild creature. This trust has allowed me to watch closely and learn things I could not learn through rare and chance encounters. Since my encounter with the tomcat, I have learned a lot about manners. I have learned that relationships must be cultivated.

~

Much of the scientific literature that describes other species is the result of studies in which the subjects are captured, tranquilized, and collared. Where such studies help us conserve habitat and understand the needs of our fellow travelers, they are important. As a naturalist, I have chosen to get to know my subjects on their own terms, letting them establish their own boundaries, but encouraging those boundaries to shrink. The smaller details and habits of daily life can be known only through intimacy.

After receiving many calls about animals in trouble, I pursued the training needed for a wildlife rehabilitation license. Most of the animals that rehabilitators care for are orphaned young. Their physical and emotional well-being depends on the presence of a nurturer. If they are young enough, their trust is readily given, and it is remarkable to watch a youngster's journey from helplessness in a very strange orphanage to proficient adulthood in their native country.

While the rewards of a trusting relationship can be large, the animal's well-being must be given precedence. Many animals would be cruelly served to believe that all humans are trustworthy. Any animal that ranges widely will encounter people who do not merit their trust. That is fine if you are a chickadee. Less fine if you are a bear or a coyote or a skunk. I draw from a background in natural history, animal physiology, wildlife behavior, and a healthy dose of humility when deciding how far to venture into the world of a wild creature.

In these pages I have assembled tales of the animals whose lives have touched my own. Part I contains stories from several years spent with the beavers that live in the great woods I frequent. Part II is a collection of tales from other places and times. They are, above all, love stories.

It is not possible to love all of life and be blind to its destruction. We are characters in a very dark chapter of the "master story," a chapter in which one rogue species has irrevocably altered the sweet, living layer of green and blue. The results are hellish indeed for many of the species of the disregarded nations. With climate change causing increasingly severe and unpredictable weather, with persistent toxins accumulating in even the remotest places, with nature losing ground to sprawling development, many of the other species with whom we share this planet are in trouble. Each day the losses mount. Each day the fabric is weakened.

I do not know what the world will be like when this chapter ends, when some new stability is restored. I do know that if humans are still among the characters, they will be fewer and wiser.

As a child, I was sure that when I grew up I would find a soapbox somewhere, stand up on it, and explain to a rapt humanity how things must be done in the future.

Ah, youth. This collection, in fact, contains very little mention of the threats to nature. Instead, I hope that these stories do for you, in some small measure, what they have done for me: for time spent in wild places and with these fellow travelers has provided a restorative respite from bad news. These excursions also remind me of what is at stake. If we all do what we can, I believe there is a chance that in some distant time when the world is again whole, there might be a person like you, on this part of our sunlit globe, who will follow a brook up a wild valley, and there find a family of beavers.

PART I

The Beavers of Popple's Pond

~

O N MAY 2, A TRICKLE OF WATER worked its way through a weakening matrix of mud and sticks in the base of the dam that contained Popple's Pond. As the flow increased to a torrent, the waters of Popple's Pond plunged downstream where, a quarter mile below, the flood aroused the interest of a crew of engineers, the likely descendents of the beavers that created Popple's Pond decades ago. Although the downstream beavers had projects at hand, something in this flood roused them; they read in it a summons to restore the pond that they had abandoned some five winters before.

And so, on May 6, I found myself once again seated on the mossy shore of this familiar cove of Popple's Pond. I see the sky reflected in ripples that slice a V across the dark, still water. The source of the ripples steers a trajectory toward my seat. Once beached in the shallow water, the beaver pauses, one paw raised. I offer her my human greeting and she strolls up the bank and flops down beside me. She reaches out with her dexterous paws and sifts through the vegetation in front of her until she locates one of the rodent nuggets I have left for her. Five winters have passed since I first met Willow on this shore. Even then her jutting hip bones suggested that youth was behind her. Now as I run my hand across her cold wet back, I can feel each knob of her spine, each rib. As we relax companionably, a second beaver steams ashore and settles down in front of me. Willow's mate, Bunchberry, closes his eyes as he chews. The sun is gone and as the landscape fades to monochrome, the soundscape becomes amplified. The robins give their good night *tut-tut-tuts* from the treetops. A pickerel frog calls from the water. The chorus of amorous spring peepers reverberates. Sundew, a veteran of two winters, arrives with an excited squeaky greeting and clambers over her father's tail to sniff for nuggets. Bunchberry, sated, relinquishes his place, slips into the buoying water, dips his head under a couple of times and then climbs back ashore on the tip of the little spit of land in front of me. With his tail sticking out in front of him and his great round belly as ballast, this mountain of beaver scrubs his belly with balled fists. In the gloaming, the sedge leaves, wet from much needed rains, shine silvery around him. Mist rises.

With these three familiars around me in this most familiar of places, I feel the comfort of a cycle completed and another beginning; for five years these beavers have shared their places and lives with me, an unlikely confederate. Now, at the beginning of the sixth summer, they have returned to the place where the adventure began.

~

A high valley in the foothills of Vermont's Green Mountains cradles Popple's Pond. The beavers live in this territory by accident of birth and perhaps I do, too. I have encountered the theory that, like birds that imprint upon their mother at a critical window in their development, we humans imprint upon a place at a certain point in our development. Indeed, I have found no place that grips me more firmly than these woods where I spent my childhood. While I have tried to live in other places, I return to these rocky forested hills. Is it because they stage the seasons with such spectacle? Is it the familiar flora? These are essential elements. Another is the wildness. If this goes, so will the sense of home.

This is true for some of my favorite neighbors, too; those species that prefer to live in remote places. I include moose, bear, bobcat, and otter on this list. I reserve a special place on the list for beavers. Beavers deserve mention for two reasons; the first is that they do much to enhance the beauty and richness of wild forests. Secondly, beavers are perfectly happy to conduct their activities in proximity to humans. The trouble arises when both species have plans for the same plot of ground. Since every bit of potential beaver habitat is claimed by one or more humans, opportunities for indignation abound. Humans seldom allow beavers to re-establish wetlands near developed areas.

Beavers are also ridiculously easy to trap, and despite the fact that their pelts have no real value today, they continue to be trapped. Colonies near any road are especially vulnerable, and so, like I, beavers do best in remote places, in the company of bears, moose, bobcats, and otters.

Six winters ago I moved into a house that sits with its back to hundreds of acres of wild forest, an area that had long been one of my favorite destinations for wild experiences. I knew some of the best places were just upstream on the brook that flowed below the house. I could imagine this brook viewed from above, a silver strand among the dark forest, beaded with bright ponds, marshes, meadows, and shrublands, the result of the dogged efforts of generations of beavers. Through the process of developing an environment that suited them, the beavers had restored this valley to something of its primal condition. This mosaic of open habitats adds diversity to the otherwise unbroken forest. Plants and animals never seen in forests live in these beaver-generated landscapes. Many forest dwellers are drawn to these wetlands, too.

On my first night as a neighbor of that wild place, impatient to get into the woods, I decided to sleep out. I headed up the brook in search of an active beaver pond. It was

September and the bugs had stopped biting so I could pack just my sleeping bag. Three quarters of a mile up the stream I found a pretty beaver pond and just north of it, a fine meadow for sleeping. The next morning I stopped at the pond to look for signs of occupancy. The pond's still surface mirrored the scarlet maples and the deep blue of a September morning sky, a perfect reflection save for the wedge of ripples that followed a beaver's prow. As I crouched and watched the beaver, I heard a heavy even tread behind me. Who would be out so far this early? I turned to see and met the gaze of a moose. She did not seem alarmed, but nevertheless continued her amble past me with greater stealth. I thought the moose and beaver encounters auspicious. This would be a place where I could be at home.

My interest in beavers might have remained one of distant admiration had I not discovered Dorothy Richard's *Beaversprite*. The book describes Dorothy's forty-year relationship with a pair of beavers and their progeny. The story begins in the 1930s, when Samson and Delilah, as Dorothy called them, were released in a stream on her property as part of the effort to restore New York's beaver population. Dorothy gave little thought to the beavers for the first few months, but when she finally hiked to the part of the property where they'd been resettled, she was astonished to find a pond populated by herons, ducks, and frogs, as well as the beavers.

The beavers altered more than the landscape, for Dorothy's life changed that day. She began to spend so much time at the pond (bearing poplar branches and apples) that she became accepted as a member of the beaver colony. My favorite photograph in the book shows Dorothy reclining pondside amidst ten portly beavers, two of them on her lap.

Dorothy's beaver affair led her to obtain a permit to raise a couple of kits and keep them in her farmhouse. She and her husband created a pond in their basement and they soon had beavers scrambling up and down the cellar stairs. Over the years, several generations of beavers occupied the increasingly elaborate accommodations the Richards provided.

I've always admired beavers, but it wasn't until I read *Beaversprite* that it occurred to me that a beaver might be an excellent companion. The book's description of the behavior of the house beavers revealed them to be creatures of great affection and intelligence—excellent qualities in a companion.

I read the book at the beginning of winter in the new house and resolved that for the New Year I would try to get to know the beaver I had seen that September morning.

Familiarity

~

I MADE MY FIRST TRIP to the beaver pond on the evening of April 15, splashing my way there in rubber boots as the remaining heaps of speckled corn snow collapsed into pools and rivulets. Once at the pond, I found a spot on the eastern shore where the water receded gently from a mossy opening, an ideal spot for picnicking and watching the activities along the western shore and at the dam.

Dorothy Richards spent her first months of beaver watching in a hidden spot some distance from the pond. I decided upon a different approach. I would behave in as non-threatening a manner as I could. Things that threaten peer silently from behind trees or skulk through thickets. Harmless creatures, I reasoned, go about their activities in the open. Harmless creatures are calm, attentive to plants, and not particularly interested in other animals. I would, therefore, sit in a visible place, make pleasant conversation, admire the vegetation, and be discreet in my observation of the beavers.

The first beaver appeared as I unpacked. He swam at a leisurely rate down the center of the pond toward the dam until he noticed me. Most wild mammals make themselves scarce once they detect the presence of a human. Not so beavers. This one changed course and did a slow pass by my seat about fifteen feet away. I said hello in a quiet, cheerful voice and pretended interest in other things. Once past me, the beaver plunged, his huge webbed hind feet sticking straight in the air for a moment, then his tail came down with a resounding smack as he disappeared. A trail of bubbles allowed me to track his underwater course until he re-emerged near the dam. Having delivered a message, he could proceed with the business of inspecting the dam.

The sound of chewing directed my attention to a second beaver on the opposite shore. While few families of any species conform exactly to textbook accounts, a typical beaver colony consists of a mated pair and their offspring from the preceding two years. Unless these two beavers had just started their lives together, there would likely be more than two of them.

When the first beaver completed his dam tour, he clambered up the far bank and pushed some fresh debris onto a scent mound, the beaver territorial sign post, gave it a couple of pats with his paws, and then waddled over the summit. Half way across he paused and did a little beaver dance waving his rump and tail from side to side. I would later note, when this activity occurred in my proximity, a complex, spicy fragrance arose.

The second beaver had finished her stick and swam ashore to choose another. To my surprise, she selected a spruce twig, nipped it from its branch and ate it, needles and all. Spruce is not featured as a preferred food on any beaver diet list I had seen. On the contrary, it is generally believed to be unpalatable. It later occurred to me that the aromatics in spruce and fir had a similar quality to the castoreum and anal gland secretions beavers use in scent marking. Indeed, I have since learned that beavers are able to isolate many plant compounds in their castor glands—often those very chemicals plants have developed to deter herbivores. Each beaver then produces a signature blend of these compounds in their castoreum.

~

Although the woodfrogs, often the first to gather and announce the demise of winter, were still quiet, a Canada goose had arrived and set up housekeeping on the flank of the weathered beaver lodge. Her mate stood, alert, on a floating log about thirty feet from me. Since the beavers were occupied on the far side of the pond, I decided to test my beaver wooing strategy on the gander. I glanced casually in his direction, informed him that he was a handsome fellow, and then busied myself with my picnic. As I gazed (idly) around the pond a couple of minutes later, I was pleased to see the wild gander tuck one leg up into his belly feathers, yawn, and begin preening.

During my first two visits, the gander, Henri, maintained a discreet distance from the nest and his mate. On the third evening he came flying up over the dam from a lower pond honking raucously. His mate stood and joined him in what seemed to be a rapid-fire call and response, *HONK-honk, HONK-honk, HONK-honk.*

She stood up, carefully covered the eggs in the down from the nest, and flew out to greet him. Following a brief interaction during which there was much flapping, bowing, and honking, Henri turned and swam straight toward me. Yes, we had exchanged pleasantries on a few occasions by then, but I was surprised when he strolled from the water ten feet from where I sat and stared at me. His mate peered nervously from behind him, but the gander just stretched his wings, yawned, and wagged his tail, the picture of goose contentment. He looked at me for several minutes before turning to follow his mate back into the pond.

I thrilled at this evidence of my wildlife communication prowess, an immodesty that Henri corrected on the next evening. No sooner had I spread my picnic than he waded from the pond and strolled over to see what was for supper. Henri had encountered hominids before. After eating his fill he stood nearby companionably and we each watched the activities of the pond at dusk.

I can't swear his attentions weren't based solely on gastronomy, but that's the way of many relationships. While I admired his black leathery feet and elegant plumage, he

admired my sprouted rye bread. I listened as the winter
wrens and hermit thrushes announced the approach of
night. I'm not sure what engaged Henri's attention once
supper was finished, but he lingered on the shore with
me. I like to think he had decided that a person made
a worthy companion. I thought a goose might be quite
acceptable as well.

~

When I resolved to make the acquaintance of a beaver,
I had little notion of the incidental rewards that would
accrue. I have always spent time in the woods, but have
generally gone in a different direction on each outing.
Now, each evening I walked the same path to my place
on the shores of Popple's Pond. Along the way I passed
the tiny spring-fed pool in which five clusters of woodfrog eggs appeared on April 20. I watched the transformation from embryos to tadpoles. As the tadpoles got
bigger, I found myself interested in how they grouped
themselves in the water. Each day I noted the arrival
or departure of birds, the advance of buds, leaves, and
flowers. I noticed the appearance and disappearance of
tracks on the trail. Familiarity allowed me to see changes
big and small.

Once I arrived at the pond, I recorded these observations in my notebook and settled down to watch the
evening. Each visit brought new treats: the snoring call

of a pickerel frog, the heron perching at the tip of the tallest snag catching the glow of sunset, the wood ducks and mergansers pausing on their search for a home, the sedge wren singing.

Each evening also brought things that could be predicted. Between seven and eight o'clock a bold junco arrived to pick through picnic crumbs. A winter wren sang from the dam after sunset. At eight I would hear the raucous arrival of the troop of grackles, their long-tailed silhouettes ghoulish against the dusky sky. One little brown bat swept across the pond at eight-thirty.

A dramatic change occurred on May 4. That morning a pair of geese flew above my house honking with agitation. Sure enough, when I arrived at the pond that evening Henri and his mate were gone. I scanned their nest with binoculars. Nesting materials were in disarray. A lone wing feather remained as possible evidence of a battle. I don't know what predator enjoyed goose eggs for breakfast. It must have been a creature that thought a swim was small inconvenience for such a meal. I thought it likely to have been the mink I watched fishing there in the winter, such a lively, graceful creature. Still, I gave little thought to the good fortune of the mink. Instead I recalled the distress I heard in the calls of the geese that morning and felt the weight of my own disappointment. I so hoped to have the company of goslings at my picnics.

On many evenings that followed, when I stayed late

enough, I heard a pair of geese fly over heading from east to west. I hoped they were Henri and his mate and that they were enjoying a summer of leisure on other ponds in the neighborhood.

~

The day the geese left I carried some striped maple branches to the pond and within minutes of placing them in the water a beaver swam over, seized one in his teeth, towed it a short distance away, and began eating. He held the stick in his front paws and gnawed the bark off in a straight, tidy row several inches long, left to right, and then, typewriter fashion, turned the branch just far enough to begin the next line and started chewing from the left again. I decided to call him Popple, since poplar, *popple* in the local vernacular, is a favorite beaver food. When he finished eating he swam back over to where I sat, climbed nonchalantly from the water, and began grooming.

That was the one of the few evenings that Popple showed any interest in my food offerings. He accepted me as a benign presence and paid no attention if I approached and sat down near him. More often I watched him from afar. He seemed to enjoy being aquatic and I could easily recognize him by the grace and frequency of his porpoising dives.

The same night that Popple came ashore, another

beaver watched from a safe distance. When Popple floated off, she swam up in slow arcs. All I could see on her approach was the top of her head. When she rolled ashore like a lumbering amphibious vehicle, her impressive beaver bulk became evident. Eyeing me nervously the entire time, she strolled to a striped maple branch within six feet of me, and began to eat the leaves. Although the front legs of a beaver can only be called stubby, she demonstrated the remarkable dexterity of her front paws; each leaf was quickly rolled into a double scroll before she fed it between her impressive chopping incisors. I would call her Willow.

I arrived at the pond between five and six o'clock most evenings. Along with branches, I brought rodent block; these sturdy little nuggets are the supplement recommended for injured or orphaned beavers at wildlife clinics. I began placing small piles of these along the shore near my picnic site. The night after Willow first came ashore to sample the striped maple, she ignored the branches and headed instead for a pile of nuggets. She seemed to locate them by weaving toward them following olfactory cues. Once she had located one, she would hold it up to her nose. After sniffing a few times, she would carry the nugget to the water and wash it before sniffing it again and then eating it. I suspected it might be the scent of my hands that worried her, and so the next night I used a scoop to handle the nuggets. Instead of leaving them in piles along the shore, I arranged them in a trail that led to my side.

When Willow waddled ashore, found a nugget and gave it the sniff test, the results pleased her. When she finished one nugget she searched out the next, often following her nose completely past it by a couple of feet before noticing the weakening scent trail and turning back. In this way she worked her way ever closer to me. I feigned disinterest and hummed pleasantly.

As digesters of the almost indigestible, cellulose-laden plant parts, beavers take chewing very seriously. Willow deliberately masticated each nugget, a process that took about half a minute. At last she arrived at my side. She glanced up at my face nervously. I said, "Hello, Willow. It's all right." This seemed to be what she wanted to hear, since she relaxed and began eating the small pile by my side.

Over my next several visits to the pond, Willow became so comfortable that she'd rest on her elbows and close her eyes while she ate beside me. If she stopped chewing and seemed nervous I only needed to speak to her in a soothing tone and she relaxed again.

I had met my goal of meeting a beaver, and by all rights could have resumed hikes to other areas or rekindled my social life. Ah, but there was the mystery of beaver number three, the rarely seen creature that I believed to be the matriarch of the colony, the beaver whose infrequent appearances suggested she was tending to kits in a nursery. And were there wood ducks in the nest box? When would the tadpoles sprout legs? Did the sedge wren find a mate?

As a lifelong watcher of nature, I felt a bit embarrassed that it took me so long to discover the rewards of watching the daily events of one place. I looked forward to becoming familiar with new things as summer advanced—firefly flashes, cricket songs, maybe even baby beavers.

As you might guess, *familiar* shares its derivation with *family*. After just ten weeks of visits to Popple's Pond, despite the mosquitoes (which appeared on May 24), the place felt like home, and the lives of its inhabitants were of consequence to me. Strangely enough, the root the two words share is the Latin *famulus*, which means servant. Perhaps that is how some of the pond residents viewed me. Especially the mosquitoes—though, come to think of it, we're probably related by blood by now.

Bug Season

≈

OVER THE YEARS, lured by the profusion of life, I have visited a number of the Earth's rainforests, where heat and humidity result in a biotic carnival. But one must pay for such pleasures, and pay in blood: plants armed with thorns and chemical irritants grow in dense tangles that impede movement; and in jungles we find ourselves a notch or two farther down the food chain, with innumerable small creatures gnawing on our flesh. Summertime visits to Popple's Pond reminded me of those tropical excursions, and I determined to suffer its inconveniences for the thrill of watching this lushest season unfold.

By June, each time I sat by Popple's Pond I found myself the nexus of a cloud of bloodthirsty insects. Their jabbing mouthparts worked their way right through denim and thick wool socks. In my attempts to deter them, I draped a second layer over my lap and feet and wore a bug net jacket and head net. Constant vigilance was needed to keep gaps from opening as I moved. Watching the pond through the fine-mesh head net was like reading with fogged glasses. Still, I grew accustomed to the incessant hum of my enthusiastic companions and felt reasonably secure, however dense their numbers. As dusk descended, and the no-see-ums arrived, my sanity wavered. Many people don't seem bothered by no-see-ums, midges as tiny as their name suggests. I am one of the sorry people in the other category. A single no-see-um bite causes intense stinging itchiness over several square inches of skin and the sensation lasts for half an hour. Even if only two or three have breached my defenses, the itching is so generalized, I believe a legion has laid siege to my flesh. Although their biting parts can't get through jeans, and the net mesh is truly fine enough to exclude them, they have a gift for locating all of the places where garments terminate; they can stroll under the elasticized cuff of my mosquito gloves as if walking through an open barn door. Their stealth attacks eventually bested all of my attempts to remain serene and I would hastily pack up, twitching with paranoia.

I suspected the only repellent that had a chance with this horde was 100% DEET. I had resolved not to go that route again after a trip to the Amazon where the DEET on my hands dissolved the coating on my camera. I began to scheme up alternatives. How about a one-piece no-see-um suit? The problem was that the little buggers could bite wherever the mesh touched skin. Hmm . . .

maybe hoops like a hoop skirt, a sort of a Michelin Man costume? Then the obvious solution occurred to me—a no-see-um mesh canopy would not only keep the bit-ers from my skin, but would allow me to wear summer clothes, move around, write, eat, drink, and bring other people with me. The downside, of course was that the canopy would be a barrier, physical and visual, between me and the beavers. Still, desperate times call for desper-ate measures.

Within a week the canopy had arrived. I carried it to the pond and strung it up between trees. Designed to cover a single bed, the canopy gave me several feet of head room and space to sit and spread out my things. Would Willow come under the canopy?

I spread some nuggets and raised the net. Willow reached her head in, looked around, and then marched in and settled down as if she had always wished for a pond-side cabana. I congratulated myself on my suc-cess. I had outwitted the slavering throng and could enjoy the singular experience of sitting under a canopy with a trusting wild beaver.

Willow finished her meal. Not one to linger and chat, she headed for the exit—that is, she turned and strode decisively to the spot where she entered. By the time I swept into position to raise the canopy, her bulk was planted on the bottom edge. As I loomed over her and tugged, Willow panicked and barged. In an instant I stood in the open air and my brilliant idea careened off through the woods to the music of snapping branches and ripping fabric. Seconds later, the net-swathed beaver plunged into the pond. My moment of stunned awe over, I plunged in after her, propelled by a nightmare vision of Willow drowning in a tangle of mosquito netting. I managed to grab the end of the retreating canopy and hold tight. Willow broke free.

I didn't expect to see her again that evening, and maybe not for weeks, but she didn't swim away. She halted her flight once she was free and swam slowly back toward the shore. As I stood wringing my hands and apologizing, Willow watched me intently, her expression inscrutable. I had never been examined in this way by a wild animal before. She stayed there for many minutes. I can't imagine what beavers think, but she gave the impression that she was deep in thought.

I didn't even notice the biting insects as I headed home with the tattered remains of my canopy. I chastised myself for having caused my beaver friend such a fright. Did she think I had tried to catch her? Would she ever be willing to trust me again?

When I returned three nights later, two beavers were swimming on the far side of the pond. While I unpacked, one of them inspected the dam, slapped its tail in warning a couple of times, and then disappeared over the dam. Damn!

As I looked around, however, I spotted Willow treading water just to the north. After a bit of coaxing she came ashore, looking just a bit apprehensive. Instead of settling in for a pleasant feed, she divided her meal into several snacks, returning to the pond in between to attend to beaver duties.

Whatever grudge she held would be of short duration it seemed. The mosquitoes, on the other hand, welcomed me back enthusiastically. As I tucked the extra shirt over my lap and pulled my hands inside the sleeves of my bug jacket, I enjoyed the general conviviality. Mind you, the no-see-ums wouldn't arrive for another hour.

Summertime

~

EACH EVENING, I continued my walks through the woods to Popple's Pond. Now instead of the melodies of birds declaring their fitness for mating, I heard the agitated chippings of parents warning me away and the begging calls of fledglings fluttering in their wake, a few threads of down still sticking clownishly out of their first suit of feathers. My walking pace slowed, and not just because of the plants that were now knee-high on parts of the path; those plants concealed tiny orange dinosaurs, eastern newts in their juvenile red eft stage.

At the pond, the beavers had changed their activities in ways that made me hope that baby beavers would soon make their debut. Willow no longer greeted me when I arrived. She and Beaver Three remained near a derelict section of dam I hoped they had renovated as a nursery. Only later in the evening, between 7:30 and 8:00, did Willow come to sit with me and enjoy the beaver treat du jour.

Dusk came later those days, and it was worth the wait. Twilight in early July cues a cast of performers that are both fascinating and highly observable. At Popple's Pond, the first of these began warming up just before dusk; one moment the thrushes were singing their lullabies, suggesting it was time for bed, in the next the air vibrated with the strident trills of revelers announcing that it was time to mate. These calls, each a sweet burst of one second's duration, were produced by the loudest and least-observed frog in these parts—the gray treefrog. Masters of the cryptic arts, their pattern of squiggled black lines on pebbly skin blends seamlessly with lichen-encrusted bark. This year, I began to hear their calls from the woods on warm evenings in May, and as the season progressed they moved ever closer to the shores of Popple's Pond.

These little frogs were not difficult to find once they had arrived at the pond, though it took a little patience. Like other frogs, they don't like to perform if they think they are being watched. Frogs provide tasty meals for many predators and calling males are very vulnerable. Still, once a chorus is going it's hard for an individual male to remain quiet. When I determined the approximate location of a frog, I moved in and waited for him to betray his position when his passion overwhelmed his sense of self-preservation.

One evening, with the chorus in full throat, I heard a duck-like squawk from the fir grove near my seat. I crept into the gloom of the thicket and the peculiar cry drew me to a spruce bough upon which two male treefrogs were engaged in a shoving match to establish which would hold this coveted wooing territory. The loser soon retreated.

Treefrogs provided the music, fireflies the visual magic. Like the treefrogs' trills, the fireflies' flashings are calculated to seduce. The males fly about blinking as attractively as they can, while the females remain on the ground or perched on a branch observing and comparing. The bigger and brighter the male's flash, the more likely he is to contribute genes that will result in brighter, flashing progeny. When a female spots such a showy male she flashes back, alerting him to her presence. This female sexual selection has led to gaudiness in males throughout the Animal Kingdom.

These flashings are also one of the ways to differentiate firefly species. Although their taxonomy remains murky, entomologists currently list about twenty-two species of fireflies in New England. The fireflies that came out early at Popple's Pond produced a single quick flash of greenish yellow light at irregular intervals. In the meadow halfway home, the fireflies produced a succession of flashes at a leisurely pace, numbering between three and seven flashes, but most often four or five. At home the fireflies flashed faster, so quickly that it was hard to count. A few flew high and maintained a single flash for a half second, making a yellow dash against the sky.

I awaited the arrival of bats with greatest suspense. I admit that vindictive sentiments toward biting insects played a role in my eagerness. Mostly however, the sight of bats assured me that they still populate our summer nights. The preceding winter brought the news of the devastating and mysterious bat die-offs in many of the caves where these little mammals hibernate.

Nine bat species spend their summers in Vermont. The little brown bat, which used to fill our evening skies hunting insects on fluttering wings, and whose nursery colonies used to fill attics, barns and belfries, is the species that died in the greatest numbers.

One evening, I saw a pale bat that fluttered moth-like above the meadow north of Popple's Pond—an eastern pipistrelle. Bats usually abound at beaver ponds. Most nights that summer I saw one bat. On a good night I would see two.

The summer of 2008 may be remembered as the summer when Vermonters became inured to thunder. Certainly, after the days of stormy weather that marked July, I was not going to let a little thunder deter me from my evening visit to Popple's Pond. When I arrived at the pond, the sky had brewed up a tower of ominous clouds. Another summer I might have turned and walked home. That night I decided to stay.

As I unfolded my camp seat and unpacked my picnic, Willow paddled over and hauled herself ashore. She soon settled down near me, resting on her elbows, nibbling the mix of rodent nuggets and grapes I set at her place.

Sure enough, the dark clouds flashed and the heavens rumbled, and soon the center of the storm dropped over the western hill and gave us front row seats. I made myself into a tent with a vinyl poncho. The expected torrent never arrived, but the tent protected me from the soft sheets of rain that swept past. The lightning snaked in dazzling flashes that turned the world first orange and then green before darkness closed around the pond again. The thunder rolled majestically above. I couldn't help but think that our human firework displays are less exciting and resolved that good thunderstorms ought to be enjoyed outdoors more often.

With the summer solstice behind us, we began the return to the dark phase of our trip around the sun, yet the season ahead still held most of the year's heat, much of the growing, much of the fruiting, and the initiation of new generations of wild life. I looked forward to seeing as much of it as I could, especially the young bats joining their mothers on the wing.

Ducky

~

A S THE SUMMER PROGRESSED, Willow became
my regular companion. She was always alert, as
wild creatures must be, and I moved slowly and spoke
softly. Her chewing was so loud that she had to pause
periodically to listen for danger. I would listen with her,
and the moment I said "sounds good to me," she relaxed and resumed chewing.

I should have been content, but two questions be-
deviled me: Did Willow share my sense of companionship, or was I just the strange animal that sat next
to her while she ate? and, Are there baby beavers? The
first question would be difficult to answer, but surely
the second was straightforward. Three beavers occupied
Popple's Pond, Willow, Popple and the mysterious beaver that I decided to name Bunchberry. Surely Bunchberry had given birth to kits and this was the reason for
her frequent absences.

According to the literature, newborn kits remain in
the lodge with an adult supervisor for the first several

weeks of their lives. The adults not engaged in babysitting bring food and bedding to the kits. Each evening Popple and Willow would make a couple of trips to the north end of the pond towing small branches and tufts of sedges. They submerged with their cargo next to the mountainous section of an old dam. I had never heard of beavers remodeling dams into living quarters, but that is just what they had done. I sometimes stood at a respectful distance and listened for the whining conversation of kits, but the whining of mosquitoes overpowered all other sounds.

One evening, when Willow sat up to sample the evening air, I had my first evidence that there were kits—Willow was lactating! She was the mother! Bunchberry must be a kit from a previous litter and had been assigned the bulk of the babysitting, at least during the evening shift.

Beavers give birth to kits sometime from April to June. The young are precocious: they are born weighing about a pound, fully furred and with their eyes open. In spite of this, they usually stay in the lodge for the first month of their life. If these kits had been born in April they would already be at least nine weeks old. If they had been born at the end of June, I might need to wait another week to see them.

As the days ticked by, doubts sprouted. Sometimes all three adult beavers were seen at once. Trips with branches continued, but less frequently. Could beavers be taking snacks to the lodge for themselves? By late July, I convinced myself that young beavers were not to be.

On August 1, I saw no sign of the beavers on Popple's Pond, so I continued upstream to the pond where they had spent the previous winter. The dam had suffered in recent heavy rains, and I found Willow busy with repairs. Instead of swimming off after finishing her snack, she shuffled around me for a while looking for other things to do. She got back into the water and floated off, but soon returned. I scratched my no-see-um bites. Willow watched with interest and began grooming herself. She then strolled back to where I had left her snack, cleaned up a few crumbs, and then waddled over to my side, sat up on her haunches, and gazed into my face. I lowered my face toward hers. She then ambled around in front of me, and repeated the procedure on the other side. She then grazed nearby for a while before returning to the pond.

This seemed like a partial answer to question number one; maybe Willow was beginning to take an interest in me. As I strolled happily downstream, I couldn't help stopping by the nursery dam. Surely any baby beavers would have revealed themselves before now. Equally surely, the beaver that swam out from behind the old dam was exceedingly small. A beaver kit! The little beaver looked at me, turned, and paddled back behind the dam. A few minutes later she floated into view again. She took another look at the strange bi-ped with the open mouth, and disappeared into the lodge. Farther down the trail I saw Bunchberry deftly debarking a branch near the pond shore. Next to her was another miniature beaver engaged in the same activity. Two kits! Life was good.

For the next two weeks Willow and Popple continued to provide most of the beaver watching activity, though I occasionally saw the little beavers from afar.

One night, Willow strolled back to the pond leaving her snack unfinished. She soon returned and a very cute miniature beaver bobbed at her side. Willow climbed right back to her snack seat, but the little beaver stopped at the shore where she commenced grazing on a salad of grasses and goldenrods. I like to think Willow was making a formal introduction. I can think of no other reason for her departure and return, but then again, I can't claim to know any mind, not even my own. Still, I allowed myself to feel honored as well as lucky as I admired the new addition to the family.

This young beaver had a pale fluffy coat that provided such buoyancy that she floated higher in the water than the adults. She reminded me as much of a duckling as a beaver, and so she became *Ducky*. When Willow left, Ducky remained in the shallow water near me. Bunchberry swam over hastily, slapped her tail, and tried to herd the little beaver away, but apparently Ducky had outgrown minding bossy siblings. I thought the little beaver made it pretty clear she wanted to stay with the animal on the bank that made such funny noises. After all her mother had told her it was alright.

The baby beavers followed different rules from the adults, though whether imposed by the parents or their own natural caution I did not know. Beavers are generally nocturnal, with the adults of Popple's Pond usually showing up outside the lodge an hour or two before dark. The kits, however, remained inside for an hour or two after dark until they were a couple of months old. As an additional precaution, they did not step ashore until they were well grown. Although Ducky would climb onto little islands in shallow water, I would not see her on the mainland until sometime in October.

The Moose that Stole
My Bicycle Helmet

~

A S THE EARTH REACHED THE POINT in its annual journey when dusk was brief and arrived ever-earlier, I returned from my evening visits to Popple's Pond after dark. To prolong the time that I could see the beavers and other pond activities, I would ride my bicycle partway to the pond. Where the trail became wet, I leaned it against a tree and walked the remaining quarter mile.

After spending many evenings in the same spot, I have noticed that certain nights seem charged with energy, and on such nights I see and hear more animal activity than usual. Heading home in the dark on one such night, I found my bicycle had been tipped over and my helmet was nowhere to be seen. Next to the bike was a fresh moose track. I enjoyed imagining that a rascally moose had trotted off with it, the chin strap looped through his jaws. I decided to head home and come back to look for the helmet in daylight. Some 70 feet up the trail, I found it.

The next day, a friend and I hiked out to work on a new bridge for a ski trail. As we started across an existing bridge, we saw a handsome, velvet-antlered moose standing in the alders upstream.

Over the years I have noticed that many different kinds of animals find certain tones of voice soothing. I don't know why that should be. Why would an animal that makes grunting noises as a friendly overture find a sing-song human voice relaxing? Still, I have begun casually experimenting to test this observation. Therefore, in the interest of science, I asked this moose, in my most comforting voice, "Are you, by any chance, the moose that walked off with my bicycle helmet?" The data records that this moose did not slink away as most do, but ambled toward us. I continued to chat (from what I considered to be a safe place on the bridge) and the moose continued to walk toward us, pausing to browse along the way. Soon he stood right next to the trail, his great head twenty feet away gazing at us with what I interpreted to be good-natured curiosity. We watched for a while more and then decided that we needed to get to work, so stepping off the bridge started toward the moose. At that point the moose opted for discretion over valor and trotted up the trail ahead of us, pausing to look back a couple of times. He soon veered off and watched us pass.

Over the course of the summer, as I returned from the pond after dark, I had heard a moose walking up the trail ahead of me, and I always talked to him. Could this be the same moose wondering what I looked like in daylight?

On my walk to the pond that evening, I noticed a couple of things that I had missed on the previous trip. The first was a bear scat that I must have ridden my bicycle past the previous night. After I parked my bike and continued along the muddy section of trail, I found the bear's tracks. They were of the right age to make the bear a suspect in the helmet heist. I examined my helmet more closely, and sure enough, there were subtle pocks that conformed to a bear's dentition. To tell the truth, I had my doubts about the moose all along. First of all, the tracks were slim evidence, there are always moose tracks on the trail. Secondly, I don't know that moose have enough rascal in them for such activities. Bears, on the other hand, are notorious rascals and are especially interested in plastic objects. Examine any plastic items you find out in the woods—chances are good that you'll find them punctured by bears' teeth.

My mind is often on bears in the autumn. This is the season when they must fatten for the winter. During these few months they will feed almost around the clock, hoping to double their weight. They need to consume the caloric equivalent of fifteen to twenty pints of Ben & Jerry's a day to meet their goal. In our region, beechnuts are the crop that makes this feat possible. Because they need to feed efficiently, and because there is little cover in a fall beech stand, bears tend to congregate and feed only in remote stands. I am in the habit of checking beech trees for bear claw marks wherever I hike. With few exceptions, I have found scarred trees only in isolated wild locations.

The summer of the helmet theft became an autumn that bears must dream about through their long winters. Not only were beech trees producing nuts, but the cherries, apples, and elderberries had outdone themselves. Such a year of bounty meant that more bears would make it through the winter in good condition, and that the pregnant females would produce cubs and probably twins (maybe even triplets). I hoped it would mean more bear shenanigans on wild nights in the summers to come.

Surprise!

~

As darkness came earlier, I often followed the beavers' activities by sound. Their chewing was so loud that I could hear them eating from across the pond. The beavers needed to be nearby, however, to hear their other noises; when Ducky or Bunchberry approached their parents, they greeted them with an emphatic series of whines, each with an upward inflection. This same whine is used by any beaver that is eating when another beaver approaches. The sound is so much the noise of infancy that I imagine they are saying, "you wouldn't take food from a hungry little baby, would you?" I have never seen one beaver take food from another. I have also seldom seen them share, though occasionally I watched Ducky snip off part of a twig an adult was eating.

During the height of summer I generally needed to get home before full darkness arrived, but as dark came earlier I would linger at the pond until twilight and became so accustomed to the route home that I found I could walk it in nearly total darkness. From the west side of the pond I'd first bash through a tangle of spruce and then I'd negotiate the hay-scented fern glade with the hidden stumps and logs. Halfway through this glade I could just make out the opening that led down to the beaver meadow and then onto the trail. There I would focus on the ribbon of sky overhead, which somehow provided enough information to keep me on course.

The second week in September brought a most blessed change: the biting insects disappeared. This ushered in a welcome possibility. I could now spend the night sleeping *al fresco* on the shores of Popple's Pond. The first night that I did so is among the most memorable. A nearly full moon lit the misty scene. I fell asleep with Willow munching next to me and Popple, Bunchberry and Ducky busying themselves at the shore. I next awoke in the early dawn to much the same scene, with the beavers grazing nearby like a herd of small buffalo. I continued to snooze and awaken until shortly before the sun rose, and found the beavers dining in peaceable assembly each time.

By late August, many of the red maples along the shore had given up working for the year and wore their celebratory red. A few leaves had decided even that was too much and lay in resplendent repose on the dark pond surface. The flowers of the steeplebush had faded and gone to seed. A lone peeper peeped. The number of tadpoles in my little spring pool along the trail had dwindled steadily all summer. When I paused in early September, close scrutiny revealed tiny legs on one of the remaining tadpoles. As I watched it, a fully formed mini woodfrog kicked across the surface of the pool, aglitter in his handsome coppery suit.

That same night I reached Popple's Pond at 6:00. I saw no beavers, so toured the upstream ponds. When I returned to my post at 6:50, Popple and Willow were adrift on the placid waters. Willow swam over eagerly, while Popple paddled off to give the dam a leisurely inspection. I had dragged a poplar branch down to the pond, and as Popple finished his rounds he swam over, nipped off a section of the branch, then turned his back on me and settled down to enjoy his snack. Twenty minutes later Bunchberry joined our party. She was still nervous around me, but the sight of her parents' casual comportment soothed her. She took a poplar branch of her own and also turned her back on me to eat. That night began her gradual warming to my presence. The party was complete when Ducky arrived at 7:35. Ducky swam first to Willow, making greeting squeaks, and then to Bunchberry and they squeaked at each other. This time Bunchberry made no effort to warn Ducky of danger.

All of this restful behavior made me just a trifle concerned. Didn't these beavers know that winter would arrive in a few short months? Wasn't the season for industry nigh?

A trip to Maine confirmed my suspicions. The beavers there had begun to stockpile branches in the water by their lodges, their larder for the winter, and had begun plastering their lodges in a fresh coat of mud. When I returned to Popples Pond in mid-October, I found no food cache and no sign of lodge improvements. Willow came regularly, and Ducky often foraged nearby, and I would sometimes see one of the other beavers. At last it occurred to me that I had been lulled into complacency by the pleasantness of enjoying bug-free evenings with Willow and Ducky. Could it be that Popple and Bunchberry had a wintering project underway elsewhere and I was missing the show?

On October 20 I saw only Ducky. On my next two visits I saw no beavers at all. Something was up!

A third of a mile upstream, my ears discovered what the beavers had been up to before my eyes did, for the stream that once meandered quietly through a beaver meadow now splashed noisily over a precipice. As I entered the meadow, I saw a brand new dam some seventy feet long. I crossed the stream and began my amazed circumnavigation of the new pond. I felt a proper fool for not discovering it earlier. I found a new lodge on the east shore, covered with mud. The beavers had already submerged a cache of branches near its entrance. The new pond, Surprise Pond was smaller than Popple's Pond, but about three times larger than the little pond where the beavers overwintered the previous year. On the east side the water had flooded a gently sloping forest, and there were several very nice places to sit and observe pond activities. On the west side an old logging road had created an opening on a steep bank that offered beautiful pond and sky viewing. I paused there and saw a beaver torpedoing beneath the water toward me. After a moment of scrutiny, Willow hauled herself ashore and lumbered up the bank to see what I'd brought for supper.

~

November settled upon Surprise Pond and gathered us up in her dark frosty cloak. For the beavers, so well insulated that they plunge into January waters as blithely as those of July, the temperature was of little significance, and when I bundled up in a down parka, hat, gloves, and sleeping bag and propped myself up in my folding camp seat, I found the chill welcome. The darkness, an old friend, brought both comfort and excitement, as the best friends do. With birds and insects silent, the small sounds of the night creatures could be heard; the loopy songs of distant coyotes, the snapping of twigs as bobcat or fox slipped past, and the small scufflings of shrews, mice and flying squirrels revealed the activities of the night forest. Bunchberry and Ducky now came ashore with Willow each night, and the sound of three beavers contentedly munching apples and nuggets joined the nocturnal soundscape.

While I love to meet the night on its own terms, if I were to continue to observe the activities of the beavers, this diurnal mammal needed some light. Because of my concern that a flashlight would worry the beavers, I first tried to peer into the gloaming with a red filter on a headlamp. The beavers swam over and asked for their apples. The next night I tried the headlamp without the filter. The beavers swam over and asked for their apples. I could see very little of what went on more than six feet from me. On the third night I tried my brightest flashlight, a 6 volt lantern. The beavers swam over and asked for their apples. Willow knocked the lantern over as she snuffled around it looking for rodent nuggets. Although I tried not to shine the light into the beavers eyes, I could not help but notice that when it happened they didn't seem to mind. They never blinked, squinted or looked away. Emboldened, I soon discovered that they are likewise unphased by camera flashes. I have found no literature on beaver eyesight, but added this observation to my growing awareness that vision is a sense that works very differently in beavers and people.

As a visually-oriented animal, I savored the beauty of the November night at Surprise Pond. The ghostly snags of long dead trees cast pale reflections on the black water, and the water in turn reflected ripples of light upon the trees. Moon and stars admired their November dazzle in the mirror of the pond.

I considered the beavers' trust a gift in its own right, but once privy to the intimacies of the beavers' lives, I had opportunities afforded to few naturalists. I could watch Bunchberry wrestle a log to the summit of the lodge and work it into place. I could follow Willow into the night forest and watch her section a felled tree into portable lengths. And how to explain the curious pleasure derived from sitting beside a rotund rodent that is absorbed in a good belly scratching?

~

On November 20, I returned from a three-day trip to find the beaver's world transformed from open water to ice. The ice offered the chance to view the beaver's activities from a fresh perspective. I surveyed their food pile, watched them enter and exit the lodge, and followed their bubble trails—all beneath my feet. I also learned about the beavers' skills as ice-breakers. I recorded three basic approaches: the pushing down vigorously with the front feet technique, the pushing up from underneath with the back technique, and the mastication technique. For the next few nights the beavers worked assiduously to maintain some open water. By November 23, however, it seemed the ice would get the best of them. Only a small hole remained. I began to make plans for re-entering human society.

After a few mild days though, the ice receded and the

beavers were back. I wasn't really ready for civilization yet, so this pleased me. Besides, I hoped for a favorable resolution to the Popple mystery. For a couple of months I had feared that a beaver was missing. Looking back through my journal, I discovered that the last time I saw all three big beavers at once was on September 5, three months earlier.

I recognize Willow and Bunchberry by a number of features. Willow can be identified by her bony hips, boxy muzzle, her protruding eyes, and the notch in her tail. Bunchberry is smaller, has a narrower tail, and the cute face and submissive demeanor of a juvenile. Popple, the patriarch, had been more difficult to inspect on land, but I once documented a double notch in his tail. When I first began to wonder about my count, I assumed that Bunchberry had moved on to find her own mate. When she began coming ashore and I could identify her, however, I started looking for Popple.

When I searched for recent Popple sightings in my notes, I found that I couldn't swear I'd seen him since September. I did find an unsettling observation recorded soon after I had seen the big beavers together for the last time. On that day a new behavior appeared; the lunge-splash. On the evening in question, Willow seemed agitated, and when Ducky approached, she launched herself into the pond with a skidding dive and disappeared. Ducky soon reappeared and demonstrated her own little lunge-splash. Later, in the dark, I could hear similar splashings repeated farther up the shore. This behavior continued for several evenings. Could a calamitous event have upset them? If so, I doubt it was Popple's demise, since in the ensuing weeks Surprise Pond materialized upstream, replete with new home and filled larder. Could the youthful Bunchberry have undertaken such an enterprise on her own? I suspect that Popple did much of the work and disappeared more recently, the victim of a logging accident or predation. I continued, however, to kindle the hope that he was working on a project in the lodge, or had developed the habit of sleeping late. I sometimes saw a beaver later in the evening who acted more aloof than the apple hounds. I couldn't quite see the tail, but maybe…

Thanksgiving served as a capstone to my seasons with the beavers. After spending a solitary hour at my usual observation post, I decided the beavers must be engaged elsewhere. I went looking and found they had felled a small hemlock into the brook. Though I saw no beavers, I did see a hemlock branch disappearing in jerks under the ice. Soon little Ducky, now about the size of a football, popped up from under the ice. I greeted her in my usual way and, to my delight, she greeted me in hers, the meet-and-greet squeaks she uses when talking to other beavers. This was a first and I was utterly charmed. I had become an honorary beaver!

The Ice Storm

⁓

L IKE ALL DENIZENS OF THE HILLS, I awoke on December 12 to find the world transformed. Everywhere were arches. Who knew crystal could be so supple? It was the noise that first conveyed the news that this was not an ordinary ice storm. The air reverberated with the cracks of breaking branches. The thundering booms of whole trees crashing to the ground added occasional variation.

The roads were impassable, so most of us had no choice but to stay home and enjoy the show. Seen from the safety of a sturdy timber-frame house, I found the show dazzling, but too remote. Before long, I could hear the Voice of Reason and the Voice of *Carpe Diem* arguing upstairs:

"You'd have to be nuts to go out on a day like this!"

"You'd have to be nuts to stay inside on a day like this!"

Like other members of my spiritual group, I resolved the dispute by asking, "What would John (Muir) do?" I headed to the beaver pond.

On that day, not only had the substance of the world changed, so had the orientation of familiar objects. The old woods road now passed over and under tree trunks and through beaded curtains that must have been branches. This spectacle of sparkle provided a striking demonstration of how evolutionary forces have shaped our trees. In *The Trees in My Forest,* Bernd Heinrich describes weighing clippings from the branches of young trees coated by an ice storm, and weighing them again once the ice had melted. He found that the birches collected much more ice than any other species. The maples and apple were mid-range, and white ash had collected the least.

Sure enough, the yellow birches along the path bowed like weeping willows. Like all birches, they have numerous slender branches. As ice begins to build up on them, they bend, and precipitation trickles down, building up thickest at the twig tips. Birches, however, are endowed with great flexibility, and will often bend rather than break. Despite this flexibility, the birches still lost more branches during the ice storm than most species.

White ash, at the opposite end of Heinrich's ice-accumulation spectrum, has relatively few twigs and they tend to point skyward like a candelabra. This reduces the surface that precipitation can land on and directs any flow to the base of twigs, the point of greatest strength. To compensate for the loss of leaf attachment sites, the trees produce compound leaves with seven to nine leaflets. Despite this adaptation, I still found I had to step over a few fallen ash branches on my walk.

The evergreen gamble of keeping needles year-round gives them the advantage of energy production on warm days throughout the year, and an energy savings since they don't need to grow a complete set of new leaves each spring. It also means the surfaces for catching and holding ice and snow are greater. Spruce and fir, trees of the north, have evolved to accommodate the snow load. Their whorls of branches grow from a single trunk. If they space things just right, snow and ice bend the branch tips down to rest on the whorl below, creating a steep-sided cone that sheds loads well. On December 12, these species were showing off the effectiveness of this strategy. The spruce and fir trees remained upright with their flexible branches bent, making a city of ice tipis among the buckled deciduous trees. The champion contortionists were the hemlocks. Their delicate young twigs and branches draped like fabric. I saw little or no evidence of broken branches beneath any of these evergreens. The large white pines in my yard, on the other hand, had dropped numerous branches, and take credit for ripping the phone and electric lines from the house. Unlike their more conservative coniferous neighbors, I suspect the white pines have invested in stoutness and rapid growth at the expense of suppleness.

It took nearly an hour of clambering to make it to

Surprise Pond, a walk that usually takes twenty minutes. The beavers seemed unimpressed by their new ice palace. The noise of rushing water and clacking branches made them nervous and kept them in the water, so I didn't linger.

Three nights later the ice disappeared. The next day I headed out to see if the trees would rebound. Those that remained rooted and intact had regained their proper posture, and I found the way blocked by only four major obstacles. I was impressed! At Surprise Pond all was calm, but messy. Branches dropped by beech and red maple littered the understory. I found Bunchberry tidying up. Like many beavers, she disdains red maple, but often eats beech, and was busy towing a beech branch to their food cache. This bounty of branches may provide a winter of feasting for the beaver family.

Many of the plants and animals of our forests have evolved to take advantage of the occasional widespread disturbance like this ice storm. I have learned that such events should not be considered natural disasters. The openings created by flood, ice, wind, and fire, allow some suppressed species to proliferate and rebuild their vigor. Such events also keep tree species on their evolutionary toes, giving individuals that fare well the chance to pass along their genes. The real problems arise when the natural part of the equation is in question, as it always must be these days.

The trees that failed the ice-load test are likely to make

the hiking and skiing less pleasant for a few years to come. Still, I will take what pleasure I can in the knowledge that the beavers and other earthbound herbivores—deer, moose, snowshoe hare—will be among the storm's beneficiaries. And I won't soon forget the beauty of those three days of ice!

Spring

~

THE ICE SEALED SURPRISE POND on December 15; the beavers' winter realm would be one of water, filtered light, and deep silence. While snug in their lodge, their world until ice-out would be small indeed.

One day in February, a day or two after a significant snowfall, I paused on a ski expedition to see if I could detect any activity. As expected, I saw no tracks or sign of feeding. I would know all was well if I found the warm beaver bodies had maintained an open vent in the snow on top of their lodge. My heart dropped when I saw a smooth dome of snow covering the lodge. What could have happened to this little beaver family? After a minute of pondering possibilities, I mustered my courage and knocked the snow from the top of the lodge. Beneath the thin surface of the dome I found a chamber lined with ice crystals—the beavers' heat hadn't melted through the fresh snow yet! When I put my ear to the top of the lodge, I could hear the faint rustlings of inhabitants.

By March 11, three months after the freezing, I skied to Surprise Pond again. After a week of melting weather, I hoped to find the beavers liberated. As I made my noisy approach, a beaver dove into a hole in the ice next to a tree trunk. Sure enough, I found evidence of free beavers where the rushing stream disappeared into the pond. I found a seat on the steep snow bank above the stream and packed down a little platform at the edge of the water, just in case. Within ten minutes a beaver emerged from beneath the ice and swam up the stream past me. I recognized Bunchberry when she surfaced. A gurgling sound announced the approach of a second beaver: Willow popped to the surface near my seat. She swam by me once and then approached. Soon she was resting on the platform, enjoying rodent nuggets. Did she dream of them during the long season of dreaming?

The platform I had packed was only big enough for one beaver. As Bunchberry tried to scrabble up the steep bank, her paws grabbing fistfuls of the loose corn snow, the bank gave way beneath her and she plopped back into the stream. After several failed attempts to gain the summit, she floated nearby and waited her turn.

I managed four more visits to the pond over the next couple of weeks, and each time I saw only Willow and Bunchberry. While I enjoyed their proximity, my concern for the other members of the beaver family grew. I continued to watch for Popple. Ducky had not appeared either. Would such a small beaver have a harder time making it through the winter on a diet of sticks?

On March 30, I determined to stay at the pond late enough so that all the beavers would be forced to reveal themselves. By then the snow had melted enough that I could sit at one of my favorite spots, an opening where the bank slopes gently into the water and I could see the dam and the lodge. Willow and Bunchie both swam right over, and I had one beaver eating on each side. They both looked comfortable and relaxed and eventually wandered off to groom and eat sticks elsewhere. The air was mild and still as the light faded. Venus emerged. A beaver appeared, no bigger than a muskrat. I didn't need any more light to recognize Ducky. Ducky! Bunchberry swam up to her, they touched noses, and then Bunchberry spun in the water and skittered across the surface in a series of quick, splashy undulations—beaver frolics! She then climbed up next to me and resumed eating. Ducky paddled off after her mother.

Popple, however, was gone. Willow would need to find a new mate, and unless a suitor arrived soon, there would be no kits this year, unless... well, unless Bunchberry was male? Since male and female beavers can't be differentiated in the field, Bunchberry could very well be male. Since the mating season is in late winter, when beavers are often isolated under the ice with their families, inbreeding is not uncommon in situations where one of a mated pair disappears.

With the new season, I anticipated with pleasure the events that would surely come and wondered what might change. How would Popple's disappearance impact this colony? Would the beavers stay at Surprise Pond or return to Popple's Pond for the summer as they did the previous year? Would Henri, the civilized gander, return with his mate and attempt another nesting on the roof of the old beaver lodge? What other creatures would appear to grace my evenings with the beavers?

How to Meet a Moose

~

AMONG MY SECRET HOPES for the new season was this one: I hoped to get to know a
moose. It wouldn't have occurred to me to entertain
such a fancy except for a report that there is a particular
time of year when a particular sort of moose is eager for
company of most any kind. The sort of moose in question is a yearling that has been in the constant company of its mother since birth. Mother moose are coura-
geous and devoted. I'd far sooner come between a bear
and its cubs than a moose and her calf. Their devotion
is time-sensitive, however, and it runs out just before
the next calf is due to arrive. The gangly yearlings are
driven off and find themselves quite alone and, according to my source, lonesome.

I would be pleased to help such a moose make the
transition to independence, but what were my chances
of being in the right place at the right moment to meet a
forsaken young moose?

The trail I follow to my rendezvous with the beavers drops for the first quarter mile through a forest of

hemlock, yellow birch, and red maple. When it reaches the floodplain of the beavers' brook it flattens, and for a stretch becomes an inviting grassy lane between low balsam fir shrubberies, shaded by a canopy of taller hardwoods. This path opens to an alder thicket beneath a wide dome of sky. On April 19, as I strolled between the balsams, a moose stepped from the alders ahead and stood broadside, blocking the path some fifty feet away. It raised its head and swiveled its giant ears in my direction. Could it be? Yes, it was clearly an adolescent moose. But was it the right time?

I stopped, greeted the moose in a friendly, quiet voice, and then turned away and feigned interest in the fir trees. The moose stood and stared. When I sat down to dig my camera out of my pack, it came a couple of steps closer and continued watching me. I took a few pictures, continued to talk, and to scrutinize the firs. The moose was a young female. Her thick gray winter coat was beginning to fall out, and sprinkled around her as she walked. She appeared to be in good condition and would likely do well on her own. Still, she looked as if a little companionship would be welcome.

The moose turned and wandered away several paces, but then came back again. After vacillating for fifteen minutes she ambled off into the alders.

What should I do? I didn't want to give the moose the impression I was pursuing her, but I didn't want to miss any opportunities for further interaction. I strolled in the direction she had headed, paralleling her path. I suspected she had headed for the brook. Sure enough, when I emerged from the woods, the moose stood in the stream about forty feet away. I sat down, said some friendly things, and looked around. Within a couple of minutes, the moose was chewing her cud and looking around, too. Often she would gaze at me, stretch her head in my direction and nod.

I have spent much of my life in the company of horses and have learned that they are excellent and expressive communicators. This moose spoke to me in that familiar language. I read the moose's gesture as, "I'm interested in you and I'm thinking about coming over, but I'm not too sure."

After twenty minutes or so, the moose wandered off again, and again I decided to follow. This time I found her standing belly deep in an old beaver pool, and she watched with no sign of alarm as I came out of the woods again. I stood on a bar matted with bleached grasses, admired a shrub, studied the woods, and talked to the moose. She behaved in much the same way, dipping her great nose in the water and splashing, sampling the sticks next to her, and looking over at me with an expression of relaxed curiosity. Finally she turned to face me and took a few steps toward me. Then to my disappointment, she remembered something else she had to do and turned and disappeared in the alders. The moose and I had spent an hour and a half together and I like to think we shared a sense of camaraderie as we enjoyed the evening on the stream.

On April 23, I found the first wildflower in bloom among the grays and browns of the Marlboro woods, the tiny golden saxifrage. Three days later the beavers' world was green, with buds unfurling and shoots extending so fast I think I could have seen them grow if I had the patience. A pair of geese was contemplating nesting on the beaver lodge where Henri set up housekeeping with his mate the previous spring. This gander was not Henri. He was shy and had a different white chin strap. The goose, however, might have been Henri's former mate. She dragged her reluctant beau over to where I sat and they climbed onto logs nearby to preen and flap their wings. At dusk a few of the vocal virtuosos—winter wren, hermit thrush, and white-throated sparrow—had begun staking claims to parts of the pond shore. The three beavers had taken care of some spring chores and were now beginning to enjoy their seasons of ease and abundance. I hoped that at least one young moose was beginning to do the same.

Stealth Beavers

≈

A FUNNY THING HAPPENS when you spend enough time with anyone reasonably likeable—a familial fondness develops. After spending so many evenings with the beavers, I found my affection had developed accordingly. Fondness is not the only emotion families evoke, of course. Many others leap readily to mind. Among them are false premonitions of doom— the tendency to leap to worst possible scenarios to explain the tardiness of cherished ones. This was, naturally, my reaction when the beavers disappeared in late June.

When the bugs came out, I set up a tent on the far side of a beautiful little wet meadow by Surprise Pond and spent the night there often so I could stay out for the late show. One night while I sat by the pond, coyotes began barking in the vicinity of my tent. At first I found this barking an amusing addition to the night music. As it grew later and the beavers didn't appear, I remembered that coyotes sometimes eat beavers. The worry seed was planted.

When no beavers appeared on my next two visits to the pond, worry's roots expanded their grip on my mind. I came early. I stayed late. Surely some tragic event had occurred to make reasonably regular rodents so alter their behavior. I told myself that my concern was not entirely irrational: coyotes, bears, and bobcats sometimes do eat beavers, though they can only attack when the beavers are on shore and they need to calculate the risks of attacking a forty-pound rodent that cuts down trees with its teeth.

On the fourth visit, I finally saw a beaver; at 9:15 P.M. Ducky paddled across the north end of the pond and disappeared into the lodge. No other beavers appeared that night, but I had evidence that at least one was still in residence.

I should confess here, that despite my concern, I also nurtured a hope. There is another reason that a beaver family's activities might be disrupted this time of year— the arrival of babies. A review of my notes from the previous summer showed no similar lapses in Willow's visits, but maybe every year is different. I knew that the chances for such a blessed event were a bit less than fifty percent, but those weren't bad odds.

When I headed to the pond on July 1, after a week with just one beaver sighting, I decided to check the one small section of stream I had not searched already, an area not far below Surprise Pond. Sure enough, where the stream once gurgled cheerfully through a constriction in the bedrock, it now tumbled over a dam. Willow floated smugly on the new pond forming behind it as if she had no idea how worried I'd been!

She beat me upstream to Surprise Pond and was underfoot in her eagerness for our picnic. I also carried in some aspen branches, Ducky's favorite treat, and as I hoped, she soon paddled over to investigate. She selected a branch and towed it off to the lodge. When Bunchberry clambered ashore, too, I enjoyed the mix of relief and embarrassment that comes when worry withers.

Our reunion took place at one of my favorite shoreline sites on a rocky outcropping that drops in steps to the pond. From there I could see most of the northern end of the pond and could look across to the lodge. This hillside was cushioned by a springy carpet of snowberry, a low creeping plant with tiny leathery leaves. From there I spent the next hour watching Bunchberry and Willow work on their lodge and listening to the dusk chorus that marked the beginning of the new month. All was well, and apparently back to normal.

At 10 P.M. I was engrossed in writing up final observations when I felt something bump my foot. I shone my headlamp into the gloom and found Willow nosing around for a bedtime snack. I obliged, and when she left, I gave the pond a final scan with my bright flashlight and binoculars. I picked up the reflection of a wake over by the lodge, and at its genesis a very small beaver. As the beaver zigzagged across the pond toward me she became, paradoxically, even smaller, as she transformed from possibly Ducky to definitely not Ducky. Welcome Snowberry!

This baby beaver, in her coat of dense fluffy fur, paddled past and I greeted her with as tight a wrap on my enthusiasm as I could muster. She demonstrated her ability to dive, and gave a comical little tail slap on the way down. Just a wee baby and already responsible for the well-being of her family.

Now I knew. Bunchberry was a male and had become Willow's mate. Could I have been mistaken about Popple's role in the colony? Anything is possible with families, but reflecting back from many years of beaver watching, I think not. In this particular colony, the young from the preceding year often spend the most time babysitting new kits, as Bunchberry had done the previous year. Bunchberry was a small beaver, at most three years old. Popple was a full-grown beaver and larger than Willow. According to all I had read, it would be very unusual for a beaver to stay with her parents into full adulthood, though not unusual for them to stay longer than the typical two years. Snowberry, herself, would remain with her parents until her fourth birthday.

Later, in my tent, I heard the murmurings of a mother porcupine and her porcupette as they worked their way across the hillside behind me. The mother made a series of low humming sounds, as if counting seconds, and then the wee one answered in her littler voice. The night had been a rich payment for my week of worry.

Two weeks later I spotted a pair of glowing lamps across the pond in my flashlight beam—eyeshine flashing from orange to green. Bobcat? The eyes watched me for forty-five minutes and disappeared only when I went to investigate. For several days after I saw no beavers. Did I worry this time? You bet. Did the beavers show up again? Every one.

Terrible Jack

~

AFTER MY ENCOUNTER with the young moose, I looked for her on many subsequent evenings. I often found tracks. Once I saw the tracks of a young moose that had gone charging off into the woods at my approach. I decided my moose had grown accustomed to its independence and had no further use for companions.

I continued to spend evenings on the shores of Surprise Pond where the three beavers amused, charmed, and puzzled me. Other creatures took advantage of the beavers' endeavors; dusk was never dull. Most nights a trio of hooded mergansers splashed down for a paddle. The whistle of air through their stiff pinions, loud and close, sounded like a jet buzzing the pond. One evening, taking advantage of the pond's remoteness from human ears, I began to sing. A barred owl swooped to a large dead tree nearby and peered at me intently. He flew off only to return a minute later to resume staring. He tilted his head, hoping for, what? Better pitch? There are critics everywhere.

On June 3, I arrived late, and found the meadow by my tentsite awash in the light of a gibbous moon. The path to the pond was in deep shadow though, and at my approach I could hear the retreat of a large animal, either a moose or a bear. I decided to circle wide to the other side of the pond. I had no reason to fear a moose or bear in the daylight when we could communicate our intentions, but an encounter in the close confines of a dark trail could result in an unfortunate misunderstanding. Somehow I miscalculated my trajectory and rather than ending near the beaver lodge, I came back to the meadow. The pale ferns and low mist seemed to glow against the black backdrop of spruce and fir. The noisemaker, a little moose, stood in silhouette in the center of the meadow. I sat down in the dew-wet ferns and chatted with him. After ten minutes or so he wandered into the shadows in the direction of my tent. I hoped he'd still be there after I visited with the beavers.

When I left the pond, a thin layer of clouds covered the moon and I couldn't see the moose. I zipped myself into the bug-free zone to enjoy the sounds of a June night. Peepers still peeped, though not with the enthusiasm they had a month ago. In the distance a saw-whet owl tooted energetically. Some small creature, light on its feet, tripped past the tent a number of times. On the hill near my tent a young moose masticated something tough with ridged molars, rattled leafy branches as he browsed, shook his head, flapping his leathery ears, and cleared his nasal passages indelicately.

A noise woke me at dawn. There, in the dewy meadow, a homely little moose met my gaze from thirty feet away. This moose, with antlers budding, was clearly a he and had made more progress toward a summer coat than the shaggy creature of April 19. He received my offering of pleasantries with an interested demeanor and then wandered off to browse nearby. I watched for awhile, and eventually got up. It was a cold dawn, so I sat at the edge of the meadow wrapped in my sleeping bag taking notes and making sketches.

After I settled myself the moose came over and stood facing me, maintaining a comfort distance of thirty feet. He chewed his cud. Yellowthroats sang in the meadowsweet. When the mountain tilted eastward enough so the sun could shine into this little valley, it was warm enough for me to get up and find out what the moose and other meadow denizens had been up to. Moose and I wandered the meadow on our separate missions. Ever aware of the other's whereabouts, we gradually moved closer together. When we paused to acknowledge each other, the moose stretched out his long snout and nodded, and I reached out my mosquito-netted arm and nodded.

The breakthrough came after about forty minutes. By this point the comfort zone had diminished to fifteen feet, at least for brief periods. I was sitting on a tussock when the moose paused in his browsing, stuck out his great snout and took three tentative strides in my direction. I rose halfway and reached out my hand. We touched each other for just a moment, and if a moose's eyes can be filled with wonder, that is what I saw there.

I did have a few more things to do, so I gradually headed back to the tent to pack up. The moose followed, and along the way we paused to *touch noses* a couple more times. These times the moose could not contain his feelings (fear? delight? exhilaration?) and whirled and bounded away a few steps after each cautious tag.

The moose remained in the meadow while I went over to my tent. Once at my tent, a grouse came toward me from the hillside, all puffed up and mewling to distract me from chicks. To my surprise, the little moose pinned its ears back, charged into the woods and chased the grouse away, not a bit impressed by her broken wing impression. I'll let you decide how to interpret that!

I was not in any hurry to hike back out to the world of keyboards and telephones and away from this world where creatures transcend the species barrier. I ambled slowly toward the homeward side of the meadow. So did the moose. I paused again to make some final notes (blackburnian warbler singing, moose eating cinnamon fern and bracken fern, ten feet away). As an *until we meet again*, the moose strolled around in front of me and extended his nose. This time hand and nose came together and apart three times.

I returned to the pond the next evening. The moose was standing by a spruce island in the middle of the meadow. I said hello and squelched across to the wooded rise where I had pitched my tent. The moose seemed pleased to see me and came right over to the edge of the woods near my campsite. When I went out to greet him, he stretched out his nose, pressed it firmly against my hand and took a big snuffle. Did this mean we were officially pals?

Imagine, if you will, the confusion such a relationship must induce in the mind of an already confused moose. He knew I wasn't his mother. Was I another kid? A different sort of grown-up? Just what sort of relationship should he have with an animal as strange as I? He decided to figure it out the way most youngsters do, by initiating rough housing. After wandering away a bit he turned and loped toward me, waving his long front legs. According to moose calf rehabilitators, this is how moose calves invite play. Even a little moose is too darned big for me to play with, and I didn't think I'd fare too well in a shoving match to establish dominance. I decided to let him know that our relationship was to be that of two mature individuals and based upon mutual respect. I conveyed this by squawking and waving my arms. The moose quickly wheeled and trotted off. Soon, however, he wandered into my campsite and began exploring. I admit to maintaining a keen awareness of the little moose's position during this interval. I could tell by the rascally glint in his eye that he wasn't satisfied by his first attempt at play, so I puzzled over a better way to establish guidelines for appropriate behavior. The next time he initiated shenanigans, I stepped forward and issued an unequivocal "No!" Strangely enough, this worked pretty well. The moose trotted off again, and I headed for the beaver pond to see what the beavers were up to.

The moose came along. He strolled out into the water a couple of times but the beavers took exception to this invasion of their pond and drove him off with slapping tails.

The next day I hiked in to Surprise Pond at noon to look for the moose (now dubbed Terrible Jack after a mouse in an A. A. Milne poem). He wasn't there. The next morning I returned to check again; still no moose. I decided to go look for him in a similar meadow about a half mile upstream. I wasn't particularly optimistic. It's a big woods, after all.

The meadow I headed for is long and convoluted, but when I approached the edge, there was Jack. He seemed relaxed. I sat on a rock in the woods by the edge of the meadow and he soon ambled over and browsed next to me. When he lay down in a bed of grass and ferns, I decided to go off and explore the meadow. I had almost returned to the napping nook when I heard the little worry noises of a moose—soft urgent humming noises surprisingly similar to the sounds made by other young animals I have known—porcupines, beavers, puppies, humans.… He settled down and started grazing as soon as he saw me and I returned to the nap meadow to retrieve my notebook. Terrible Jack had not seen me leave, however, and soon came charging through the copse separating us, crying with greater urgency. Again he settled down when he found me and soon returned to his bed and his nap. By this point I was overdue for an engagement and slipped away, leaving the dozing moose to his dreams.

I had not gone more than several hundred yards, however, when I heard the cracking of branches and the calls of Terrible Jack. At first he trailed me shyly, keeping well behind, but soon was just twenty yards behind me. Then he began leapfrogging, trotting ahead, waiting for me to pass, and then trotting ahead again.

The difficulty now would be persuading him not to follow me home. When we arrived in the meadow by Surprise Pond, however, Jack laid down for another nap, and when I tip-toed away he stayed behind.

I couldn't get back to the pond for a couple of days, but decided that might be just as well. I knew that only a brief relationship would be allowed and I should not let Jack become too attached. Moose are listed among the many creatures who should not harbor the notion that all humans are their friends.

When I arrived on the evening of June 10, Terrible Jack was standing by the little spruce island in the meadow. He appeared more self-assured. I wandered over to the edge of the pond to visit the beavers. After I had been settled for several minutes, Terrible Jack arrived. He stood quietly next to me on the other side of a little hemlock, gazing out over the pond and chewing beech leaves. Bunchberry munched nuggets on my other side. Peepers peeped.

After half an hour, Terrible Jack turned and trotted away. I haven't seen him since. I like to think he felt comfortable on his own then, and had moved off to begin the independent life of an adult moose. Though brief, my time with Terrible Jack provided something I had not known with a wild creature before—we had been drawn together by a mutual interest in friendship, with no food as a bribe. Though long separated on the journey of evolution, that ancient kinship allowed moose and human to reach across the misty meadow and forge a bond.

From the Shores of Lake Dismal

~

ON AUGUST 10, I reluctantly abandoned my beaver-watching post on the scenic shores of Surprise Pond. The beavers, revealing appalling aesthetic judgment, had moved to their new pondworks downstream, a spot I called Lake Dismal. The new pond arose where the stream had begun a dam of its own, wedging some debris across a narrow cascade.

When the gods are smiling, such a natural dam site will have a broad floodplain upstream, and will provide aquatic access to a fresh bounty of woody vegetation. In this case, the stream flowed through a narrow bowl, and most of the trees within it were spruce, hemlock, and red maple—trees generally disdained by these beavers.

It may have been the spruces that prejudiced me against their new pond—as a lifelong tree enthusiast, red spruce is the one species I regularly curse. The spruce that grew in close ranks along the new pond shore reinforced this bias. The only way to reach the pond was to thrash through a thicket of rough dead branches in the understory. The canopy of the trees concealed the sky, and only the occasional croak of a green frog suggested that any creature aside from the beavers would recognize this forested puddle as a pond. By the end of that first evening, however, I found that Lake Dismal had an enchantment of its own.

After long evenings enjoying the sky-filled vistas of Popple's Pond, and then Surprise Pond, I had forgotten my troglodyte heritage. Somewhere in my lineage must be a people with an affinity for caves, for I have always been attracted to dark enclosed spaces. Once I had settled in for my first evening at Lake Dismal, I shone my bright flashlight down the length of the pond. Several trees canted across the water, and the beavers had already removed the bark from their bases. The bookend reflections of these pale trunks in the still black water created the effect of a tunnel. A barred owl fledgling swooped so low above my head that I could feel and hear the air rush beneath her wings. She perched on a branch above my head and peered down at me, to make sure I knew that swoop had indeed been performed for my benefit. The fringe of down around her head produced the affect of a saintly halo.

As is typical when beavers move to a new site, their first home was a burrow in the streambank. In such dens, the entrance is underwater, and the beavers excavate a tunnel that leads up to a dry chamber above the water table. I have surveyed the shores of abandoned ponds and found that there were numerous beaver-generated burrows. These are places where the beavers can retreat if their primary lodge is destroyed or rendered inaccessible by ice, flood, or drought.

Unbeknownst to me, the entrance to the beavers' new home was right next to my seat. There, in the brook just below me, the beaver family emerged. Willow and Bunchberry came ashore to eat. Ducky stopped next to me on a pass downstream and scratched her belly before proceeding. She returned in a moment towing a branch from a sugar maple that had been felled into the pond. She climbed up on a stump across the stream from me and began to eat. Snowberry, who had been deep in the maple branches with her father, decided this would be a good time to pester her older sister. As she swam by I noticed a beaver of the same size right behind her. Those sneaky beavers had hidden a second kit from me all of this time! Soon the entire family had gathered in this narrow stream and the confinement of the space added to the intimacy of the night.

All of the beavers soon became very accustomed to my presence. Willow and Bunchberry often left the pond to greet me when I began floundering through the spruces. The need to step over beavers while ducking branches made the pond approach even more challenging. These two beavers were now so relaxed that they would lean against me while they ate.

The two babies (I called them both *Snowberry* since I couldn't tell them apart) did not yet leave the security of aquatic environs, but they soon became expert at bobbing for the apples I tossed into the water for them. This was no easy feat, as all who are old enough to have participated in an old-fashioned apple bob can imagine. It's true, the beavers are allowed to use their hands, but when your paws barely reach your chin they can't be considered a tremendous asset. During their first exciting encounters with apples, they would swim up and leap on them. If they managed to submerge the apple in the process, they would dive in pursuit. The apple would pop up again, while bubble trails marked the kits underwater efforts to locate the runaway loot. They finally learned to sneak up on the apples, extend their stubby paws cautiously to either side, and then with a quick grab, pull it within biting range.

The only time I saw them leave the water was when they climbed into the branches of trees felled into the pond. They pretended an interest in the edible parts of the tree, chewing bark as they clambered about. They didn't fool me with their *I'm a serious beaver* act, though. Once, one of the Snowberries climbed onto a horizontal trunk that one of the adults was chewing. The trunk bounced with the efforts of the big beaver and the roly poly baby straddled it as if it were a mechanical bull. She was bucked into the pond three times and each time scrambled back into the saddle.

Once the bugs had stopped biting, I could sleep beneath the spruces on the shores of Lake Dismal. The beavers remained active all night, and Willow sometimes woke me as she clambered over my legs on a quest for stray beaver treats. Sometimes I would hear something rummaging around nearer my head, and would turn on the light to find Ducky beside me asking for an apple.

In mid-September a dome of sticks and mud began to rise steadily on the bank above their burrow. Peering into this thick latticework with my flashlight one night, I saw one of the Snowberries, glossy and dry, peering back at me; the beavers had dug up through the ceiling of their burrow and had expanded their living quarters into this new lodge. My sleeping location was next to one of the beaver runways, short paths that they use to build up speed as they worked on their home renovation project. Throughout the night I heard the patter of webbed feet as the beavers, embracing loads of sticks and mud and rising onto their hind feet, hustled as far up the dome as they could before collapsing. Once the cargo was deposited they pushed it into place with their paws, rocking forward and back to take full advantage of their bulk. Larger sticks were worked into the structure one by one and later chewed off from the inside as the beavers shaped their chambers.

As autumn progressed, I looked forward to watching the lodge and winter food cache grow, the water level rise. True, the stars would be few and the vistas uninspiring, but I had no worries that evenings at Lake Dismal would be dull.

Things that Go Bounce in the Night and Other Small Wonders

~

THE GREEN FROG SAT PERFECTLY STILL, only his white throat moving in and out with each breath proof that he was not a figurine. We have all seen green frogs. We have admired their gold-flecked eyes, the mottled variety of colors on their glistening skins. What made this frog unusual was his evident interest in *me*.

Green frogs are one of this region's two predominately aquatic frogs, but unlike bull frogs, they do most of their hunting on shore. I first noticed the frog when I heard a hop in the woods behind me. Each time I turned to look, he had hopped closer, until he simply sat for fifteen minutes a few feet behind me. I wasn't surprised to hear the *ker-plonk* as he eventually hopped back into the stream. I was surprised to turn a little later to find him sitting right beside me again. And there he sat, a garden statue, for the half hour I remained at the pond.

When he returned the next night and sat for as long as I did, I hope you will excuse me for feeling just a bit special. Unlike the beavers, who sat next to me because I brought them food, the frog sat there for reasons I could not fathom. He showed no interest in the insects lured to my light.

On the following night there were two frogs and the night after that, there were three. They stared at me as if mesmerized. I began to fancy they mistook me for a goddess from the green frog pantheon. I suspect the truth is they were drawn to my flashlight, but I thought it quite an amusing spectacle—those little green buddhas meditating upon me.

~

It had been a summer of wonders small and large. While I am always thrilled to see a moose or a bear, there is much to recommend a subject whose prosaic business occurs within an area so small you can see much of it from one spot. I met one such creature while exploring an abandoned beaver pond upstream from Surprise Pond. From across the small pond I saw an animal about the size of a mole crossing the top of the old beaver lodge. I circled the pond and settled near the lodge to watch. The first thing I noticed were fans of tiny ripples along the shore, and then the source of the disturbance revealed itself—a small black ball-shaped mammal skittering along on top of the water. She then leapt from the surface of the water to the top of the bank. The leap was only about six inches, but I found this defiance of physics astonishing.

Once she was ashore, I recognized her body plan as *shrew,* and based upon her performance I knew I had my first sighting of the elusive northern water shrew. As I watched, this shrew worked her way around the entire pond in this fashion, alternating forays ashore with saintly perambulations across the water. At last she reached the lodge again. As she busied herself within three feet of where I sat, she looked like a black velveteen ping-pong

ball with a long pointed nose at one end and a longer tail at the other. Oblivious to my scrutiny, the shrew continued her manic search for insects, inserting her mobile snout into every crevice before disappearing, finally, into a tiny hole in the surface of the beaver lodge.

I have since learned that water shrews frequent mountain streams where they take advantage of the abundance of aquatic and terrestrial insects. Their buoyancy is so great that it is only by kicking vigorously that they are able to dive down to explore the crannies of a streambed. As soon as they stop kicking they pop back up. Their jaunts across the water are made possible by a combination of this extreme buoyancy and by stiff hairs on the sides of their feet that trap air bubbles beneath them.

⁓

On a few nighttime walks home from the pond I have been baffled by footsteps in the woods nearby—the irregular tread of a single light foot—step…step, step……… step…step, step… These footsteps did not fall in logical proximity, but as if their maker rode a pogo stick. One night I heard this sound near my tent. My flashlight revealed a mouse leaping in high, erratic bounces—a woodland jumping mouse. I wondered how the little creature could ever get anywhere, but after several seconds of this crazy behavior, he settled down and scooted off into the shadows like an ordinary mouse. I suspect that this behavior is an effective predator baffle.

I knew very little about jumping mice, and had seen them only rarely, most often bounding quickly across the road in my headlights. I knew of them as members of the elite club of *deep hibernators*. Now I also know them to be the nocturnal pogo stick clan.

One night at Surprise Pond, I noticed that the beavers were not the only rodents enjoying the snacks I had brought. In the gloom I saw what appeared to be a nugget attempting to lurch away from an impending beaver. In my flashlight beam I saw that attached to the leading edge of the runaway nugget was a creature only about triple the size of said foodstuff. In a moment this red-backed vole had maneuvered himself behind the nugget and managed to transport it quickly, if not gracefully, away.

The next night I brought a handful of sunflower seeds to the pond with me. I scattered these in the mossy bowl beneath the little fir trees by my seat. The first diner, a little mouse, appeared as soon as she deemed the dark deep enough. I am *not* one of the very few people able to reliably distinguish deer mice and white-footed mice in the field. Both members of the genus *Peromyscus*, these mice have large eyes and ears and neat white bellies and feet. To complicate matters, the characteristics that can be used to distinguish them in one locale do not apply across their ranges. I will choose to call this one a deer mouse. She sat quite near me and daintily shelled each seed before stuffing it into a cheek pouch. I didn't know that deer mice transported seeds in cheek pouches. While this creature's cheeks did not achieve the proportions a chipmunk's can, there were obvious bulges before the mouse scurried up a nearby fir tree and made off through the branches. This also solved another of the many nature mysteries I have filed away: a friend once reported finding a neat little pile of shelled sunflower seeds under each pillow on his bed. He kept only sunflower seeds in their shells on the premises. The good fairy must have been a *Peromyscus* mouse.

Soon more mice appeared. One moved quickly and nervously, stuffing whole seeds into her cheeks. Only a few seeds could fit in this manner, and she soon bounded away, her face stretched like a cartoon clown's. Others practiced a mixed transport strategy, shelling some seeds and pushing others in whole.

I had no idea how large my fan club had become! Soon a woodland jumping mouse slipped into the mossy amphitheater and began eating discreetly beneath a fir bough. As hibernators, jumping mice do not cache food, but do need to get fat. I had plenty of time to admire this bold little mouse. Her shiny warm-brown coat had a band of darker brown that stretched from her forehead to her tail. Her tail disappeared partially into the moss, but I managed to trace it to its white-tipped end. I had to check three times to be sure it was really all tail. I have never seen an animal with longer tail-to-body ratio.

Another jumping mouse arrived. And another. Whenever they bumped into one other as they searched for seeds, both would leap into the air. Their landing would startle the others and they would all leap, and for a few moments the glade would erupt like popping corn. It seemed their bouncings lacked design, since they often bumped into me and landed amidst my gear.

The next night I carefully placed the seeds in little piles to minimize the chaos. How I enjoyed the scene of the peaceable kingdom—the one red-backed vole trundling back and forth transporting one seed at a time to his larder, the deer mice neatly shelling and stuffing, or just stuffing and departing, and the pretty little jumping mice sitting at their piles and eating their fill. Even the beavers seemed interested in the increased activity. While they snuffled about looking for rodent nuggets, they would sometimes stop and eye their tiny cousins with interest.

When the beavers moved their activities to Lake Dismal, I was sorry that I would have to leave the mouse show to follow them. I brought sunflower seeds to my new seat, of course. I expected some mice might find them eventually, if I returned to the site often enough.

Once darkness arrived I heard something behind me. My first mouse already? I turned and found the handsome green frog, his golden eye glittering in my flashlight beam.

The next hopping sound marked the arrival of a

jumping mouse. Two jumping mice and two deer mice found their way in the dark of night to this rich new food source within an hour of my arrival. I suppose I shouldn't have been surprised. Locating seeds is critical to their economy. I couldn't help but admire sensory acumen so superior to my own.

I suspect that such small wonders might be seen on any night almost anywhere on the planet if one ventures forth with a pocketful of sunflower seeds, or perhaps some resemblance to an amphibious deity.

Let it Snow

~

DECEMBER IS THE MONTH I begin my snow
incantations in earnest; snow can't come soon
enough for me—assuming I have my firewood in, the
snow tires on, and hay for the goats. I awoke on December first to find a light dusting of snow. Everything was
ready except the snow tires. That was good enough for
me. The gray squirrel family I had been watching since
September seemed to be prepared, too. The four young
siblings had moved into a cavity in a large yellow birch
at the edge of the woods. I had watched them clambering among the delicate branches of the trees outside
my window, stuffing their mouths with leaves for their
winter nest.

For the past three months they had secreted away
the chestnuts, walnuts, and acorns I gathered to sup-

plement their scant local resources. Each hard shell encased a treasure of fat and protein. With nuts in their jaws, the squirrels bounded elegantly across the yard, their tails towed behind like banners. They pondered one spot and then another, ever aware of who might be watching.

Once a site was decided upon, the little paws pushed aside the surface debris and dug a tiny hole, then, hind end aloft, the squirrels pushed their prizes into the ground with their muzzles. Finally, paws spread like a toddler playing patty cake, they arranged any available litter over the hole and earnestly patted it down.

The gray squirrels had good reason to be surreptitious. Not only were their siblings scurrilous thieves, but the treetops were filled with pirates, their spyglasses raised. Soon blue jays and red squirrels would swing from their crows' nests and begin their looting. How many times each nut was handled before it reached a final resting place, none can tell.

The beavers continued to work steadily. Their winter preparations on December first included a pond deep enough to avoid freezing to the bottom, a canal system, a fine big house plastered in mud that would freeze as hard as concrete, and a winter's supply of branches stored just outside their door. They had accomplished all of this with only a set of chisel teeth, little hands, stubby legs, and occasional assistance from webbed feet and big bellies.

Their slow-but-steady approach to tasks served them well, but the season was not without adversity. One evening in mid-October, I stopped to visit and found Willow too nervous to come ashore. None of the other beavers appeared. I hiked back the next night with a couple of friends. Willow did come up to join us, but her snacking was punctuated by pauses to emit huffs of agitation. She made frequent trips to the pond and back, and once trundled over and did the waggle dance beavers do when they are marking scent mounds. Sure enough, a sweet perfume arose from the spot when Willow stormed back to the water. Did the scent marking suggest that a strange beaver had passed through their territory? I hoped that was the case.

I brought the same visitors back five days later. This time the beavers had resumed their normal behavior. When we sat down on the shore next to the lodge, Willow and Bunchberry settled down next us to enjoy a picnic of rodent nuggets while Ducky delicately plucked apples from my fingers and carried them to the water to eat. One of the Snowberries swam back and forth nearby waiting for us to toss her apples. She had become expert at wrestling floating apples, and when she subdued one, she swam into the lodge to eat it. Each time she entered the lodge, I listened for the apple-defense squeaks that would prove that Snowberry II was in the lodge as well. Silence reigned.

After two weeks of seeing just one baby, I accepted

the evidence. An unfortunate event had occurred. The babies were about the size of footballs at that time, and had just begun tentative investigations of the shore. Coyotes, bears, and bobcats were on my suspect list, with coyotes at the top since I often saw their tracks and scat in the area. I love coyotes, those beautiful, wild, golden dogs, but I wish they had a different way to make a living. I know this to be the bargain of life in nature—beauty and abundance cannot exist without ruthlessness and strict economy. At least in nature the beauty and wonder are omnipresent, and even a sad night on the shores of Lake Dismal contains wonder.

That sad night, as I reflected upon the short life of the Snowberry, the nose of a short-tailed shrew provided diversion. Short-tailed shrews often betray their presence with tiny, musical squeaks when they discover sunflower seeds. On this evening the shrew had discovered a small pile of sunflower seeds sixteen inches from its tunnel entrance. After scurrying back and forth to grab seeds, it remained in its tunnel. The ground between the tunnel entrance and the seed pile began to tremble, and I heard the sound of teeth severing little roots as the shrew homed in on the seed pile from below. Soon a mobile pink snout emerged in the midst of the seeds. With growing confidence, the entire sleek, shiny mini-mammal zipped into view, grabbed a seed, and hustled in reverse back underground. Unlike moles, with their great digging paws, the ever-speedy shrew somehow managed to tunnel to the seeds using its tiny front paws and teeth, and presumably its nose for navigation.

Each time I visited throughout the fall, the pond was larger. My former sleeping spot by the lodge was now underwater. The flooded area extended out into an old beaver meadow, and enough of the canopy trees had been felled that I could see some stars from my seat beneath the spruces. The pond would need a new name if it became any more impressive.

Their lodge had become a mountainous igloo of sticks and mud packed by the passage of many big bellies and webbed feet. When the walls freeze, the beavers will be secure from predators and even the worst of winter weather.

In late October the beavers began working on their winter food cache. They piled branches of yellow birch, beech, maple, and hemlock in the deepest channel in the pond, jamming them into the muddy bottom to anchor them, and wedging later additions into the tangled mat. With four fat rodents to feed, it probably won't be enough to keep them growing all winter, but with luck it will be enough to keep them from shrinking. Northern beavers become less active and have considerably reduced caloric requirements in the winter and by the end of November these beavers had entered their lowered metabolism phase. With their interest in food diminished, I was lucky if they spent five minutes visiting with me, but this meant I spent more time watching them go about their beaver business. One important piece of business was especially fun to watch. Beavers need to keep their fur carefully groomed and oiled to maintain its thermal and water-repellency attributes. Bunchberry often settled down near me after eating to attend to this task. He would plant himself on his rear end with his tail and webbed feet sticking out in front of him, and would bend forward and give his stomach a good rubbing with both of his front paws, massaging in the waterproofing oils from glands near his tail. He would give his face and ears the same treatment, and then comb his back and sides with his immense webbed hind feet.

Once Bunchberry finished grooming, he would float off to see what needed to be done. He always found something. Near as I could tell, though, these four rodents had all their autumn chores done.

Sample Size

~

MARCH THAWS LIBERATED the beavers the
next spring. With the snow still two-feet deep,
they were able to enjoy food that had been out of reach
when their pond froze. Wood chips festooned the muddy snow by their plunge holes. I was pleased to find
the beavers of Lake Dismal fat and healthy after their
months of seclusion.

Two years into my beaver-watching project, I was beginning to think I had learned a great deal about beavers.
What if I had only learned a great deal about these particular beavers? I decided to find another beaver family
for comparison. I had seen beavers at two other locations in the watershed. In both cases, the dam and lodge

were uninspired, and on return trips the beaver was no longer there. Last December, a friend and I discovered a new dam and lodge in the brook directly below my house. I looked forward to spring and a chance to have a beaver so close to home. Alas, by spring the site had been vacated. I searched upstream and down and found no fresh beaver sign.

In mid-April, I decided to take the scenic route to visit the beavers at Lake Dismal, that is, to follow the brook the whole way instead of taking the trail. I hoped to find a sign that bears had been feeding in the wetlands. I found no bear sign, but I did find a set of very fresh beaver tracks. The next day I returned to see if I could find where the new beaver had settled. With so many shrubby meadows to choose among, I wondered how a beaver would decide. With each new track, scent mound, and chewed twig, I imagined that surely the beaver had chosen a location. The next few days I returned, expecting to find additional signs of beaver musings, but found no fresh sign of the peripatetic beaver. I had to widen my search.

It took me eleven days to find the beaver. He had chosen to settle on a tributary well above the main brook where a small dam repair had recreated a pretty pond. A fine new lodge had been constructed on the shore. Newts and minnows now floated above the carpet of grasses, sedges, and dewberry brambles that were so recently a meadow. As I marveled, the proprietor emerged from his lodge and swam past to inspect me. Could it be that this entire pond and lodge had appeared in the eleven days since I'd seen the fresh beaver sign on the main brook? I have learned not to underestimate what a rotund rodent with stubby legs, webbed feet, and a flat tail can accomplish.

On my second official visit to the pond, a happy discovery helped explain how so much work had been accomplished; two beavers were in residence! I would have a proper family for my comparison project. While one beaver slapped his tail before swimming off (all reference to gender is random at this stage), the other swam over to a little island right in front of me, about thirty feet away, turned her back to me, and began grazing. She then climbed ashore, turned to face me, and started to groom herself, rubbing her marvelous belly vigorously with both front paws. Finally, she gathered a mouthful of sedges, plopped into the water, and carried her cargo of bedding to the lodge.

Their new pond was a very satisfying place to sit as the sky darkened and the diurnal creatures traded shifts with the nocturnal. Winter wren and hermit thrush, with dueling melodies, ended their day, while barred owl and saw whet owl began. The peepers, like a manic hand-bell choir, generated a swelling din. At the height of their performance they paid no heed if spotlights shone, thus I walked among them to enjoy the visual performance, too. Whirligig beetles, their eyeshine headlights alight, circled each other on the surface of the pond like bumper cars.

After two weeks, my status with the beavers began to transition from *curiosity* to *welcome benefactor*. They showed no interest in the apples and rodent nuggets at first, but the poplar boughs always disappeared between visits. Soon, one of the beavers decided to inspect my offering of sugar maple saplings while I sat about twenty feet away. She swam close, then lost her nerve, turned and slapped her tail. I have learned to gauge the seriousness of a tail slap by how long the beaver remains submerged afterward. This beaver popped right back up and headed for the branches again. I talked to her in a quiet, cheerful voice. This time she strode ashore, tugged a branch free from the others and swam ten feet farther up the shore before settling down to eat. Alas, it was late and I needed to leave. I would have to stand and walk toward the beaver to get to a break in the trees. She would likely consider this threatening and abandon her post, but it couldn't be helped. To my pleasure, she remained at her meal. I left the pond to the serenade of peepers and a beaver's noisy gnawing.

Within a few weeks, the pair would swim over when I arrived, for they had developed a keen interest in apples. One of the pair delayed gratification, however. Though apples bobbed nearby, he floated, his nose just a few inches from the shore, and stared at me. Beavers have a limited range of facial expressions. They don't smile or frown or prick up their ears. When I say that his expression was inscrutable I'm not saying much. Perhaps it is a beaver's habit to be still and gather sensory information in unusual situations. Perhaps he actively wondered about me and my intentions. Perhaps I was a curiosity to be marveled at. I will never know. Eventually his trance would be broken and he would poke about for an apple.

By May, Lake Dismal had indeed outgrown its name, for the pond spread into a pleasant meadow across from the lodge. On a typical night I would arrive at the meadow and be greeted by a small stampede of beavers. Willow, Bunchberry, and Ducky welcomed me as the bearer of exotic delicacies. Snowberry remained more circumspect, but just as eager. She came ashore when I arrived, but then retreated to the pond once I sat down, and would await apples delivered from the heavens.

Apples from heaven present a curious difficulty for beavers. Because their diet is composed mainly of large, stationary objects, I theorized that beavers have not honed the skills that help other animals locate food. The food I bring is quite small, and seems to make no visual impression upon them at all; they seem to rely entirely on olfactory information to find my treats. What's more, the beaver nose is apparently an imprecise organ that leads them back and forth, slowly homing in on the food. You would think an apple tossed conspicuously to a spot within a few feet of the intended recipient would send out a visual beacon. Not so. Snowberry would note the splash and would swim toward the apple, then past the apple, turn around and end up on top of the apple, and would often give up the search, returning to gaze beseechingly at the purveyor of fruits.

That spring I decided that, like Ducky, she should learn to take apples from my hand. What she learned instead was that apples could be reliably located near the distal end of the large stationary object that sometimes sat on the shore. She learned to stalk carefully toward me until she reached my feet. Once there she stopped, sat up, and waited until I bumped the apple slice against her nose. After a second's pause, she grasped the apple in her teeth, and then waddled briskly back to the safety of the water.

~

After spending so much time in the company of beavers, I might be excused for assuming a small level of expertise. This would not be tolerated in scientific circles where I would need statistically significant data to support my hypotheses. Indeed, later that summer I met a beaver who did not fit my theoretical model regarding beavers' ability to locate food.

I met this beaver while exploring three miles downstream from Lake Dismal. This was beaver heaven—acres of alders and sedge meadows in a wide valley. I sat down on an old section of dam across from what appeared to be the beaver habitation. Beautiful green fireflies flickered among the alders, a white-throated sparrow sang its lament, green frogs and gray treefrogs called, and at last a beaver appeared from around a bend.

A friend and I made several trips to visit this beaver. Although Mike (never ask friends what you should name a beaver) was wary of us, he readily developed a taste for apples. Here's the strange thing: Mike could swim across the pond straight toward a floating apple, grab it, and carry it off, no correction of course necessary. Is Mike an exceptional beaver, or are all of the other beavers I know handicapped in some way? My modified hypothesis now is *most beavers have difficulty pinpointing the location of small delicacies, even aromatic ones.*

Other creatures have no such problems. As has become my habit, I carried a few sunflower seeds with me on my visits to the new beaver pair. One night, as the ponderous beaver and I studied each other, I heard the bouncing approach of another creature in search of food. Soon the sound stopped, and I saw the first jumping mouse of the evening busily shelling the sunflower seeds I had placed a couple of feet from my seat.

The approach of the jumping mice is amusing. They arrive in evasion mode—high, erratic bounces designed to foil a predator. One mouse ended up in the pond several times last night. I wasn't surprised, therefore, when the mouse hopped again and landed on my lap. I was surprised, however, when its next little deliberate hop carried it directly into the plastic bag of seeds. It settled down inside the bag, not bothered in the least by the giant illuminated creature watching it. Do I dare say that jumping mice are remarkably adept at locating food? I have learned my lesson and will make no assertions without more evidence.

Willow

~

DURING THE THIRD SUMMER with the beavers, I decided to offer my services as a groom. Since beavers spend a great deal of time with their ablutions, I hoped such overtures would not be unwelcome. During the first spring of beaver watching, Willow informed me that advancing hands were construed as threatening. Valuing my digits and our relationship, I kept my hands to myself. Given what we have learned about each other in the ensuing seasons, I now felt confident that any reprimand would come in the form of a huff of warning or an indignant retreat. While Willow sat next to me and ate, I lightly draped my arm along her back. She continued to eat. I lifted the arm and resettled it a few times. Her reaction was the same. I then allowed my fingers to burrow a bit into her damp coat. No response. I made so bold as to stroke her back. She seemed indifferent. Indifference is not the preferred response when intimacies are offered. Still, I considered it a good start.

When Willow did not show up for the picnic for two evenings, I hoped her excuses would be fluffy and flat-tailed. This third season I hoped to have earned privileges denied me in previous years. During my first summer of beaver watching, when I sat down within fifty feet of the nursery lodge, Willow bustled over huffing and blowing, and almost dancing in agitation. I felt fairly sure she was suggesting I find another place to sit, and I did so. Last June, I watched baby-season activities from a discreet location across the pond from the nursery. This year, given the trust bestowed upon me by the beaver parents, I had every expectation that I would be permitted to sit wherever I liked to await the kits' debut.

Like most baby animals, beaver kits are irresistibly appealing. Even the most stalwart curmudgeon of my acquaintance remarked, "That is a cute baby beaver!" when I forced him to look at one. It surprises me, therefore, that Willow does not appear to dote on them as I would. While I cupped my ears and listened for telltale squeaks from within the lodge, Willow sat beside me and nibbled rodent nuggets, floated around the pond, or worked on construction projects. I seldom saw her enter the lodge. Snowberry seldom came out.

In early July, my impatience was rewarded when a pair of brand new miniature beavers paddled into view. I welcomed the small creatures and heaped lavish praise upon their parents. The parents, resting on their elbows, lost in gustatory reverie, ignored kits and me. The little beavers swam to the shore and floated there, gazing at the peculiar scene—the bulky backsides of their parents flanking a strange, noisy animal. After marveling for several minutes, they discovered the poplar branches I had placed in the water. With their tails curled up out of the water for balance, they floated and chewed.

The bank I watched from was the remnant of a little meadow that then lay beneath the waters of the pond. When crossing that meadow a year earlier, I nearly stepped on a snowshoe hare that had a run beneath the dewberry brambles. Dewberry, I thought, would be an appropriate name for a beaver born at this pond, and since I couldn't tell the two kits apart, they became the Dewberries.

The night that I met the kits, Willow returned for a second helping of rodent nuggets. She swam past her children without greeting them, but one of them followed her ashore and peered coyly at me over her mother's back. This surprised me, since the beaver kits from the two preceding years did not venture ashore until the early autumn. I assumed they stayed in the water for safety, but did their parents teach them this behavior or was it baby beaver common sense? Willow did not reprimand the kit for her incautious adventure. This bold Dewberry returned to the water, but when Willow left again, both kits swam over to the shore and rose onto their haunches to look at me. After a moment, both waddled out of the water. I reached my hand out toward them, and one walked up to within a few inches before turning and bouncing back to the wet.

I strolled home that night with delight and concern wrestling for control of my thoughts. Were these babies too foolish to survive? Unless kits stay in the water, they are easy prey for a number of predators. On the other hand, I had been developing a trusting relationship with this beaver family for three years. Maybe the kits' behavior was simply an outgrowth of that trust, and not a cause for concern.

Delight prevailed for the next couple of visits as the kits continued to thrive and entertain. Two weeks later, concern entered my thoughts again. Willow, usually the first beaver to swim up when I arrived, did not appear. Bunchberry, Snowberry, and one of the Dewberries kept me company. Although it was unusual for Willow to miss snack time, I assumed she had things to do elsewhere. Still, I stayed until 11 P.M hoping to see the missing mother and kit. As I began walking home and neared the mighty Lake Dismal Dam, a beaver lunged into the water and slapped with her tail— Willow. I waited for her to come up so I could reassure her. She did not appear. This was strange. Concern romped unhampered through my mind on the walk home.

On my next visit a few days later, Snowberry and Bunchberry showed up first, and then a Dewberry paddled up with Willow behind her. Willow did not come ashore right away, but swam in wide arcs around me for several minutes, the kit clinging to her tail. When she did come up, Dewberry followed, but soon returned to the water. Willow, never as fastidious as her sleek mate, seemed more disheveled than usual. When I shone a flashlight into her fur from behind, I discovered four parallel lacerations. The only thing that would produce such injuries through a beaver's dense coat was a bobcat. The missing Dewberry did not reappear. Since her injuries occurred at the time the kit disappeared, I think it likely that Willow attempted to defend her kit from the bobcat. While I can only imagine what she felt, such an act requires valor.

Now, three weeks later, the remaining Dewberry behaves like a sensible little beaver. Instead of sitting in the shallow water to eat, she tugs a poplar sprig into the lodge to dine. Willow's wounds have nearly healed. She rests on her belly beside me, eyes half closed, chewing diligently. She may be immune to the allure of cute, but I no longer question her devotion to her kits.

The Castor Master

⁓

SOMETIME DURING THE TWO DAYS following
the birth of the kits, Ducky disappeared; the arrival
of a new generation had apparently cued little Ducky to
begin the dangerous adventure of independence. While
staking claim to needed resources is a challenge for every young animal, beavers face more impediments than
most because of their need for water. What's more, they
can not settle in just any wet place; rivers that are too
big to dam must also be deep enough that they won't
freeze in the winter. Streams that are small enough to
dam must have a gradient shallow enough that the dam
will form a pond, and the pond must provide access to
adequate food. They must also be outside of the territory of established beaver colonies. Such places are few,
but are made far fewer by the territorial claims of another
species—*Homo sapiens.*

The low-gradient streams, streams with floodplains
suitable for damming, also happen to be the easiest
routes for human travel and the best places for human

settlement. Beavers that attempt to reestablish wetlands in areas humans prefer to keep dry, or to chop down trees that humans value, tend to have short lives. Even when their activities cause no damage, the closer a beaver pond is to a road, the more vulnerable its occupants are to trapping. Beavers being beavers, they seldom occupy roadside habitat without acquiring the wrath of road crews. Culverts provide beavers with a nice narrow channel to dam; ponds can be created with ease! Road crews clear culvert dams only to find the dam restored overnight. Removal of the beavers was once the only solution, and then just a temporary one, for such sites will continue to attract footloose beavers.

Could I help to keep Ducky safe in such a hazardous world? If she ended up settling near humans and their roads, I might need to intervene on her behalf. I would need to broaden my defense capabilities. I knew who could help.

Skip Lisle, the person who has done the most to help beavers, lives several towns north of me. I drove up for a visit and to learn more about his work and his company, Beaver Deceivers International. From the center of Grafton I followed Hinkley Brook Road up a hillside to the place where Cabell Road splits off. On the short drive from the beginning of Cabell Road to

Skip's house I traveled along a plateau that has become a wildlife Shangri-la, thanks to beavers and Skip.

The backyard of Skip's house looks over a large lush pondscape. Before we headed out to explore it, Skip brought out a photograph—a framed panoramic view of the pond taken in 1970. The photo showed the sort of pond that humans build to improve *their* environment. A grassy lawn surrounded the shore, and short steep banks dropped to the water. The pond was just what Skip's parents wanted, but offered little to the other residents of the area.

Skip's father moved his family to Grafton in the 1940s because of one attraction—the grouse hunting. Grouse proliferate in early successional and shrub habitat of the sort that spread over Vermont's then recently-abandoned farmland. Grouse are not the only species to thrive in such habitat. Woodcock, white-tailed deer, bear, moose, cottontail, snowshoe hare and certain songbirds all benefit from the young vegetation and fruit crops generated in these sunny places. These shrublands are now one of Vermont's rare habitats.

Twenty years ago, Skip told me, he moved back to Grafton, and back into his childhood home. He had a very different landscaping paradigm from that imposed by his parents. As we strolled across his modest strip of lawn, he marveled at the time and energy Americans spend manicuring their yards, and in the process, destroying habitat. Skip chose to allow natural succession and a team of engineering rodents to take over the bulk of his landscape maintenance. The result is wildlife habitat as rich as any I have ever seen. Since quality wildlife habitat is the landscape Skip values the most, his plan succeeded brilliantly.

The path to the pond entered a wide swathe of shrubland. After a very wet summer, the fruits of his strategy were evident. We stopped to admire a laden choke cherry, one of many in the mix of young trees. Though bird activity was waning in late summer, Skip pointed out a cedar waxwing sitting on her nest in a small white pine, and the mewling of a catbird. Catbird numbers are a good measure of the quality of shrub habitat, Skip noted. A bit closer to the pond we began to see trails the beavers made through the grasses and shrubs.

Unlike the somewhat reserved vegetation that grows in the acid soils surrounding Popple's Pond, the vegetation in the Grafton complex was a lush extravagance of shapes, textures and colors.

When we paused to admire the pond, I asked Skip how his interest in beavers developed. He admitted that when the first beavers moved into his parents' pristine pond, his father told him to shoot them. He shot two, and found the activity so distasteful that he refused to shoot any more. Over the ensuing years, the pond began to change, and in ways that pleased the younger Mr. Lisle. For a start, the pond grew larger and soon the steep banks were underwater. Lush wetland vegetation began to grow in the shallow water along the widening shoreline. Some trees were drowned leaving snags that became habitat for cavity-nesting birds.

In the 1980s, Skip left a career in construction to head to Maine for a graduate degree in wildlife biology. He studied beavers, naturally. When he finished school, the Penobscot Nation hired him to help solve their beaver problems. The Penobscots' lands are like much of the rest of interior of Maine—largely wet and flat, with many clear-cuts growing up with poplar, the favorite beaver food. Beaver populations were robust, making it very difficult for the people to maintain roads. Shooting the beavers didn't seem to help, as young beavers filled vacancies quickly. Skip worked with the Penobscot for six years, and over that time he developed and perfected the flow control devices he called *beaver deceivers,* a name now used widely for all similar structures. The task was daunting, with the beavers only too happy to point out the flaws in his designs. Skip told me that because of his commitment to beavers and the ecological and hydrological benefits their presence brings to a landscape, he refused to fail.

Over the years Skip has tested materials for durability, aesthetics, and their ability to thwart beavers.

While each situation requires its own design, Skip uses two basic structures in most of his installations. To prevent beaver dams from clogging culverts, heavy-gauge wire cages force the beavers to dam well back from the culvert. In addition, a large plastic pipe, its intake located nearer the middle of the pond, drains off any water above its set level. Thus water flows freely through the culverts, and the pond level can not rise above the level of the drain pipe.

What a pleasure to stroll through the complex of habitats created by Skip's beavers, and with someone who appreciated them at least as much as I did. We followed the path around to The Terraces, a pretty series of stepped dams that the beavers have been working on over the past decade, through a blackberry thicket, and up to the road. Here we paused to admire one of the beaver deceivers that keeps the Grafton road crew happy.

At the far side of this wetland Skip showed me how he protects trees and shrubs that he doesn't want the beavers to remove. A simple fence of eight gauge ungalvanized mesh two feet high surrounded an azalea. This wire is rigid enough that it can just be pushed into the ground. No stakes are needed to hold it up.

Our tour of Skip's beaver lands drew to a close in a grove of tall aspen adjacent to a small wetland full of willows. What could make a beaver happier? The beavers had begun visiting the area. Skip was pleased, but wasn't sure he wanted all of the aspens to come down. That presented no problem, though. If he decided to keep any of those trees he'd just fence them.

When Skip returned to Grafton from Maine, he became a member of the town selectboard. He noted wryly what an uncomfortable fit public office can be for those

of us more at home in swamps than sitting around tables. Still, discomfort is sometimes necessary to safe-
guard nature. In Skip's tenure on the board he made the
town of Grafton a haven for beavers by educating public
officials about the value of beaver-generated wetlands
and by installing his Castor Masters on all of the town's
vulnerable culverts.

~

I returned home fortified. If Ducky settled in hostile territory, I could make practical arguments for harmonious
coexistence. A few alterations could regulate the depth
of the pond and simple fences could protect any selected trees. The best arguments might best be made by
taking skeptics for a short drive. There are many places
along the back roads of this region where beavers have
made ponds in the past, only to be removed. In two
places, however, the beavers have been welcomed and
protected. In each place, a stately white farmhouse looks
over a valley that features a magnificent beaver pond.
The shores of these ponds are grazed each spring by
gaggles of goslings. Friends who canoe on one of these
ponds report encounters with otters. Whenever I drive
by them, I stop to watch for wood ducks, kingfishers,
herons, and beavers. If such ponds become a regular
part of our roadside scenery one day, thanks will be due
to the determination and perseverance of Skip Lisle.

Ducky Adrift

~

WHILE THE NEW KITS SPRAWLED on fresh bedding, secure in the Lake Dismal Lodge, Ducky faced the big wild world alone. I guessed that she would follow the brook and its tributaries rather than head off cross country, though I had once seen a beaver dead in a road far from any stream, so I couldn't be sure. Would she look for a mate first, or suitable habitat? A mate would be difficult to come by. I knew of no other colonies that would have dispersing two-year-olds.

I found no sign of Ducky upstream from Lake Dismal. Nor were there new impoundments between Lake Dismal and Mike's wetland at the southernmost end of the brook. Two tributaries flowed into the brook. My new beaver pair occupied of one of these. If Ducky had not settled on the other, she must have left the watershed and moved out into the big river. With a road following that river for most of its length, and with many tributaries she might have wandered up, Ducky would be much more difficult to find if she had left the watershed of her natal brook, and much more exposed to the hazards that arise from proximity to humans.

On Sunday, July 11, I set out up the remaining tributary. I knew that beavers had once occupied a site about three-quarters of a mile up, and a pretty little pond had persisted long enough to appear as a landmark on a few generations of maps. In the midst of this hot, dry summer, the cool water and deep shade of the little valley were most welcome. There were no signs that a beaver had passed that way recently—no chewed branches, scent mounds, or nibbled vegetation—until I approached the upper reaches of the stream. Did this small dam have some fresh mud packed along the top? Around the next corner all doubts disappeared. Here were fresh damworks and evidence of recent meals. At the pond I sat to eat my lunch on a grassy bank beneath some pines and then headed east along the top of the dam. Fresh mud along the top and sprigs of blueberry in the water nearby suggested the pond had at least one resident beaver. It was early afternoon so I didn't expect to see beavers out. I clambered noisily up a bank and into the woods at the end of the dam. As I did, I heard a beaver surge through the water and slap its tail. A trail of mud and bubbles revealed that the beaver had exited a derelict lodge built long ago on the bank. I sat down near this lodge and scanned the water. On the far side of the pond, I saw a beaver float nervously out from behind a bleached snag. Ducky? I called to her. The beaver did not respond with alarm, but began to paddle slowly toward me. I tossed an apple into the water. The beaver paddled faster. Ducky!

Ducky located the apple and floated in front of me to eat it. Though I hoped for a prolonged and joyful re-union, Ducky ate only part of the apple. She paused in the entrance to the lodge to groom and then headed back to bed.

I returned to visit Ducky several times in the weeks that followed, always at a more civilized hour. Though I often made considerable noise bushwhack-ing through the drought-dry woods, Ducky was never again alarmed by my approach. Instead, she would be waiting to greet me by the shore and would come right up to collect her apples.

The habitat was better than that at Lake Dismal, but not excellent. Like most beaver ponds at this elevation, red maple, spruce and fir ringed the waters of the main pond. A patch of alders grew in one section, though, and young woody growth surrounded the pools she had enhanced downstream. There were also some pond lil-ies, which are eaten by beavers, and their roots provide winter sustenance. The occasional beech and yellow birch would add to her stock of reserve food.

Perhaps she settled there simply because the pond occupies the headwaters of the first tributary down-stream from Popple's Pond. She may have worked her way up the stream as far as she could and set up housekeeping when she could go no farther. This location had one important amenity that made it superior to Mike's amazing alder swamp downstream; the nearest human disturbance was more than a mile away in any direction. Ducky's activities would not offend any humans.

She lived in the pond alone, though, and I wondered how that would affect the life of a social animal. How would she find a mate? Would she wait for a dispersing youngster to find her or would she go on a quest sometime in the future?

I imagined her solitary life there to be a bit gloomy, so I was not entirely surprised when she disappeared again a month later. I renewed my search. I visited the pond at the confluence of the river. Mike remained the solitary occupant. My next step was to follow the length of the brook from my house up to her childhood territory.

As I approached an old dam, I saw that water had collected behind it. Sure enough, a beaver's wake sliced across the dark pool as I approached. Ducky swam right up for her apples. This site did have an abundance of winter forage, and though just a fifteen-minute stroll from my house, it was still remote from other humans. Beaver trails, canals, and a harvest of alder branches suggested that Ducky was not just passing through. As I gazed upstream, admiring the pretty valley, I saw the second beaver. He swam right over. Ducky, busy with an apple, greeted him with a few little squeak-whines. The new beaver emitted a low growly hum, then a warning huff, and for good measure, slapped his tail. When he surfaced, I explained that I'd known Ducky since she was in pigtails, so he'd better get used to my visits. He huffed again and swam off.

Despite this cool reception, I was delighted to meet Ducky's mate. In the fairy tales this is where the story would end, and I could conclude with "and they all lived happily ever after." I preferred to think of this as the beginning of a story, and so it was to prove.

High Water

~

STREAMS DRIED TO TRICKLES, leaves drooped
on their twigs, and as rainless summer day followed
rainless summer day, muddy shores expanded while
ponds shrank. As the summer progressed through July
into August, and then on to September, the beavers had
no dam work to do. On September 26 it began to rain.
By the time it stopped four days later, the beaver ponds
were full and the streams were high.

A friend and I set out to find out how the beavers
would respond to this abundance of opportunity. We
stopped first at the homestead of Ducky and Growler.
After Ducky came over for her apples, Wilson and I set
off for Lake Dismal, a half-mile upstream. At Lake Dismal, though water thundered over the dam, there was a
strange stillness—no mob of eager beavers greeted us.
Upstream from the din I called to them, and at last a little beaver paddled down the stream. Dewberry, the kit,
appeared to have been left behind while the others went
off to celebrate the bounty.

On my next visit, two days later, I followed the stream to Lake Dismal again. I found ample evidence that beavers had been on the move. Fresh mounds of mud and vegetation had been pushed up in many places and were redolent of beaver. I put some of this mud in a plastic bag to see how the Lake Dismal beavers would respond. Did it belong to one of them or to an outsider like Growler?

I reached Lake Dismal at dusk, and all four beavers greeted me. They ignored my scent mound sample, and so I knew it was theirs. Soon they were settled in and re-laxed, enjoying the picnic. A few mice scampered about busily, stashing sunflower seeds. Wood frogs clucked from the woods nearby, restored by the rain. A porcu-pine wandered in the dark, whining to itself. When the beavers finished their snack, the adults returned to their logging operation on the far shore. Dewberry strolled be-hind me to rummage through my pack, walked around to the front and gave my boot a taste, and then sat up, put her paws in my lap and looked into my face. No more apples?

All had returned to normal. The only sign of the beavers' adventure was a hole in Bunchberry's cheek, just the size one would expect from a beaver bite. Sev-eral nights passed before I could return to Lake Dismal. When I did, only the mice appeared.

The next morning I headed out to look for any sign of beavers. The large dam just upstream from Ducky's pond had recent improvements. As I sloshed through the wet meadow behind it, I saw a beaver in the stream channel. Ducky? When the beaver paddled over, I recognized Dewberry! The Lake Dismal clan had staked claim to the finest habitat in the brook. Ducky and Growler were gone.

On my next visit, I encountered the yearling Snowberry at the new pond. She came ashore and huffed, then charged at me and stamped her feet. I held out an apple. She took it and, continuing to blow and glare at me, she carried it to the water to eat. The confused yearling seemed to be in warrior mode. As a member of the conquering clan, perhaps she had been expected to drive her own sister away. Did she see me as another familiar but no-longer-welcome community member?

I have read that beavers recognize close relations by scent even if they haven't met before. According to the beaver literature, these family members are welcomed. Not so outsiders. Had Ducky and Growler teamed up to defend their pond? Bunchberry's battle wounds (I found another injury behind his shoulder) suggest that Growler, at least, tried to guard their territory.

Until that day I had imagined that this brook could accommodate three or four beaver colonies. No more. Here's why: suitable territory is a strict limiting factor in the beaver economy. Unless beavers live in a large river or lake, they are restricted to streams with narrow parameters for gradient and flow. Food resources along such streams are also limited. The success of a beaver colony depends on maintaining a territory large enough to meet their needs over many years. My beaver colony seems to move at least once a year. This allows shrubs to grow up in their abandoned wetlands, shrubs that will provide food one day when the beavers return. In this sense they are gardeners and would be foolish to let foreign beavers settle in the areas they will need in the future.

The size of beaver territories is partly governed by the quality of available habitat. Beaver densities will be much higher in areas where their preferred foods proliferate. In the low-quality habitat available to Willow's clan, I have found that they mark and defend a one mile stretch of brook. As with other species, the number of young produced is linked with the quality of habitat. Willow has had two kits each year. Dorothy Richards carted loads of poplar to her New York beavers every day. Delilah, the matriarch of her clan, sometimes gave birth to six kits. She certainly did her part to repopulate the Adirondacks!

～

From the shores of the big new pond I watched the November full moon rise through the naked trees on the far ridge. The beavers had been making dramatic entrances, breaking up from the water through a cascade of shattering ice. Willow and Bunchberry then settled beside me and calmly chewed beaver nuggets. They let me dig my fingers into the dense silky undercoat that keeps them warm and dry in the frigid water. Dewberry and Snow-berry ate halved apples in the water nearby. They came ashore for refills. Snowberry touched the toe of my boot with her nose and then waited for delivery. Dewberry strode up and put her paws on my lap. Behind them, in the middle of the pond, I could see their refurbished lodge and the top of their food cache. They were ready for winter in their new pond.

I had not relocated Ducky, although I extended my search into the neighboring drainages. In the process I found that the beavers from two of the only three other occupied sites in the area are also gone. I hoped that the high water just got everyone moving, and that I would eventually find all settled in safe places and provisioned for winter.

Source or Sink

~

ECOLOGISTS USE THE TERMS *source* and *sink* when describing the ability of different areas to support a particular species. In a source area, a species not only reproduces, but creates surplus population that then moves out into surrounding areas. A sink, as the name suggests, is an area where these same animals would vanish altogether without overflow from the source areas. Because an animal is seen in a sink area, we might imagine that the area is supporting a healthy population. But if the source disappears, the species will disappear from the sink.

For the six months since Ducky's disappearance I had been contemplating the sources and sinks for beavers. By my crude beaver census of several towns in our region, it appeared that when beavers move to areas adjacent to roads and houses, few survived long enough to reproduce. Only the few suitable streams in remote valleys provide habitat where beavers can live full lives.

This gloomy pondering was further fuelled by the

disappearance of Mike from his bit of beaver paradise at the mouth of the brook, a paradise that lay quite close to two roads. His dam and lodge were far enough from the roads, and in such impenetrable alders, that I hoped he might be undetected by trappers. When I discovered him missing, I found recent evidence of a small-scale horticultural enterprise. I hoped that Mike had not been shot to protect a marijuana nursery. Old beaver meadows seem to be a preferred location for such activities. The beavers upstream had ignored a similar nursery, so I had some reason to believe that beavers do not find *cannabis* palatable. Still, that is not something I expect a grower to know. That same day, as I worked my way upstream through the alders, the disconcerting distant drone of ATVs became a roar—the machines could only have been coming right down the middle of the brook! Above the roar I could hear shrieks and laughter, followed by revving engines, and eventually cursing. They had made it down to the beaver impoundment and become mired in silt. I could not see them from my position, but listened for half an hour as they attempted, unsuccessfully, to free the machine. I was glad that I did not find Ducky there.

My search for Ducky became easier with the arrival of snow and ice, for following brooks and exploring old beaver meadows on skis is a joy. Excitement kindled as I approached each old beaver meadow. As I crossed one stream after another off my list, my hopes of finding the two beavers diminished. At the end of my list of likely prospects was a brook near the headwaters of the big river. I knew large beaver meadows would be found about a mile up the brook, and I set off on skis feeling hopeful. My hope seem warranted when I spotted a snow-covered dome at the far end of the meadow—a beaver lodge! As I approached, however, the stream continued to meander in a most un-pondlike fashion and the lodge became the roots of an overturned tree.

The return ski along the brook was just as lovely as it had been on the way up, but I felt overwhelmed by disappointment. Ducky was the first beaver kit I had met. I had spent many hours in her company, watching her learn and grow. If she were unsuccessful in finding a territory of her own, what hope did her younger siblings have? In March, I decided to ski up a stream on my list of Unlikely Prospects. I had been disappointed so often that I told myself I was just exploring. Although the stream was a tributary of Ducky's childhood brook, I considered it too small to be attractive to beavers. As I neared the headwaters, I could make out a meadow beyond the dark hemlock woods. I checked my excitement. At the edge of the meadow, however, I saw freshly chewed beaver sticks in the water. Ducky? I noticed a hole in the ice and dense layers of sticks in the water. Beside me, some beaver-chewed poles poked out of a hummock at the base of a tree. I was standing on an occupied beaver lodge! I skied a dozen feet from the opening in the ice, unpacked the apples I had brought, just in case, and called to Ducky. Soon the surface of the water in the little hole began to ripple. The ripples grew! A beaver's head popped up. The beaver scrambled eagerly onto the ice and grabbed the apple I had tossed toward the hole. Ducky!

March played its seasonal havoc. Most days the ice was too thick for the beavers to emerge. On one spectacular blue-sky day, however, I made the ski to Ducky's refuge. An inch of fresh snow had returned winter to the throne. The preceding days had been seasonably warm, however, and a large hole had opened in the ice by the beaver lodge again. I brought apples and some striped maple branches. I called to Ducky and waited. Again the water began to roil, and Ducky emerged from the depths. This time she came and plucked an apple from my fingers and waddled to the edge of the ice to eat. After two more slices, she nipped off a section of branch, dove into the water and disappeared. A few seconds later, the surface rippled as another beaver surfaced. The beaver muzzle that appeared this time was not the pretty, youthful face of Ducky, but the boxier muzzle of an older beaver. Growler, Ducky's mate, had not become accustomed to me during the few weeks I had visited with them before their eviction. I expected he would quickly slap his tail and disappear. Instead, he gazed at me for a moment and then swam over, found an apple, and turned his back on me and began to eat as he floated. He then nipped off a branch and chewed off the bark, still floating just eight feet from my seat.

My relief at finding Ducky was tempered by the awareness that the pond would not sustain them for long. Although the location was remote there was not enough forage for beavers. I thought it a wonder that the pair made it through the winter. When they moved on, would they be able to choose a location in a source area? In this landscape with its many human claims, the beavers' time-tested discrimination no longer applies.

This spring, if all goes as anticipated, Ducky will have kits. So will the new pair of beavers that settled in the watershed last spring. With Ducky's parents, that makes three sets of kits. When I think how few places there are for the youngsters to safely settle, I fret. Still, how lucky these beavers are to occupy the source, and how lucky for those who value wetland habitat that sources exist.

Spontaneous Generation and Other Dog Day Occurrences

~

THERE HAD BEEN A STRING OF THEM, those dog days that suck all of the cool from summer, and I guess it was on a dog night that I poured a pail of brook water into the parched depression—my own experiment in spontaneous generation. In my flashlight beam I watched as the disturbed silt settled and air bubbles rose. I had little enough reason to think it would work; the ancients believed that, while some creatures arose from mud, salamanders were born from fire. Still, they appeared in the puddle, sudden and whole and wriggling, as if as surprised as I to find themselves recalled to life.

It's true, I had reason to hope for this particular magic. Three dog nights before I had seen the gilled spotted salamander larvae concentrated in the tiny remnant of water, all that remained of the small wet pool where their misguided mother had laid her cluster of eggs, just a logging rut, not a proper vernal pool at all. Despite such folly, the great damp of June had propelled a batch of miniature wood frogs from the pool's modest depths, and these two-inch salamanders were within a few weeks of following in their pondmates footsteps. Surely, however, the pool would cease to be during the morning hours of the impending dog day.

I am always tempted to intervene in these situations, to give doomed lives a second chance and to avert suffering. This temptation usually wins after a brief argument with the inner ecologist, the one that reminds me that salamanders with foolish parents should be removed from the gene pool. In this instance, I had to give way to the scientist since it was almost midnight, the nearest bucket was at my house, a quarter mile away, and I needed to leave at dawn for a three-day trip.

And so it was after an interval of three dog days and three dog nights that I returned with my pail. The pool had well and truly vanished. Still, once I poured the water onto the earth, there they were, the tiny spotted salamanders, as if generated from the mud itself, I herded them onto my submerged hand, transferred them to another pail of brook water and gave them their second chance.

To discover how other creatures were managing the heat, I decided to spend the next afternoon in the woods dangling my feet in a brook, for the sake of science. When I heard the beavers slapping their tails I set off upstream to investigate. When I neared their pond, I found a bull moose splashing his way past their new lodge, the lodge that likely sheltered the new beaver kits of the year. The moose's antlers had achieved about two-thirds of their growth and were still in velvet, but his majesty was somewhat diminished by the sounds he was making. How to describe them . . . a series of wheedling grunts? For a moment I thought he was talking to me, but then I heard the splashing of another moose upstream. He continued toward the object of his affection, clambering easily over one of the beavers' new dams along the way. I thought it little wonder the beavers were disgruntled.

I think of moose as winter beasts, masters of deep snow and frigid temperatures. When summer temperatures are above 80°F for extended periods, they lose weight and can go into winter with insufficient fat reserves. I wasn't surprised to find these moose cooling off in the beaver pond. I listened for a quarter of an hour as they splashed in the shallows. When the moose disappeared into the thick growth on the far side of the stream, the beaver agitation intensified, and this time it was directed at me. Bunchberry swam past me, paused, and huffed menacingly before diving and slapping the water with his tail. I did not take offense; I knew he was reacting to the satchel I carried, the one I call the porta-possum. More specifically, he was reacting to its contents, three small fluffy opossum orphans. I thought it remarkable that a beaver living in a remote valley in the Green Mountain foothills should have formed such a strong opinion about a species he had likely never encountered before.

I decided to hike farther upstream hoping to find the moose cooling off at the next pond. To avoid offending the beavers, I chose a route well up on the hillside and out of sight. Bunchberry followed along in the stream below, however, uttering his foulest imprecations. I decided that was enough science for one dog day and took the possums home.

That night I returned to visit the beavers, as free from *eau de opossum* as I could manage. When Bunchberry appeared he had a deep gash on his front leg. While the opossums might have roused his suspicion, he had other reasons to be upset. Had he been injured in an encounter with trespassing beavers or a predator? I would later learn that sometime during that period, Ducky and Growler moved through the territory of Bunchberry and Willow, and this was the most probable cause of the excitement.

The dog days waned, as they will, and the logging rut filled with water again. I decided to return the salamanders to their source, knowing that if the dog days returned, I could perform a little alchemy and raise them again from the dust.

Irene

~

VERMONTERS SELDOM WITNESS the full brunt of
Nature's majestic disdain for the ambitions of her
creatures. It is not to be wondered at that so many of us
were unprepared for the heavenly deliverance of a southern sea; Irene was, after all, a tropical storm. Nature may
not consider the fate of her creatures, but as I watched
the water from my windows, I certainly did. I wondered
especially about the wild neighbors with whom I have a
relationship; Priscilla the squirrel, in her leafy drey nest,
must be getting drenched. Charles, the woodchuck that
has a burrow beneath a rock in my meadow, might be
protected from the flood. What about the chipmunks?
Surely many denizens of the earth would be driven from
their burrows by the rain. I was especially curious about
my beaver colony. Would their dams withstand the high
water? Would their lodges flood? As the water reached its
highest level in this little watershed, I set out into the rain
to find out.

To say that I was unprepared for what I saw when I
reached the beavers' brook is an understatement. Where

the beavers' modest stream once flowed five feet below a skidder bridge, it now roared over the top, a seething torrent fifty feet wide. How would the aquatic insects, salamanders, and fish fare in this overwhelming deluge? Their nurturing, buoyant medium had been transformed into a smashing, grinding, driving flood; their sheltering rocks plucked and tumbled by water rendered powerful by sheer volume.

As I worked my way upstream, water flowed in sheets through the woods, and I forded knee-deep streams where I had never seen streams before. Over the course of the summer the beavers had constructed a brand new series of dams as well as a new lodge. They had flooded an alder thicket that would provide good winter forage. In this wider part of the valley, Irene's tide slowed and spread out into the forest. I waded to an island hummock that once abutted the beavers' dam. The dams were gone. I thought I could make out the top of their lodge, but I knew my chance of actually seeing any beavers was remote in this chaos of rush and rumble and the veil of falling rain.

Downstream I found that the two venerable old dams, reinforced by many generations of beavers, appeared largely intact, though nearly invisible in the landscape flattened by the flood.

Later that afternoon the sun came out. Charles the woodchuck emerged from beneath his rock and grazed. A chipmunk, also dry, sat up on another rock. Crickets and grasshoppers creaked out their same songs. From their vantage and mine, the world seemed unchanged. For those whose lives were connected more directly to the water, however, I would need to take another look.

Once the water returned to the sea, I explored the brook again. My notion of how the world is formed began to change. I had fancied that floodplain landscapes are shaped gradually through erosion and deposition. At the skidder bridge I found the course of the stream had changed. A bar of mid-sized stones now blocked the former course of the brook and sent it to the east a bit. The little pool from which I had resurrected the spotted salamander larvae lay buried beneath a pile of cobbles, the bed of a stream that lived for just a few hours. I hoped the salamanders had completed their metamorphosis in time.

The root mat of the streamside vegetation had been peeled and rolled back, revealing the tunnels of shrews and voles in the naked mud. How many residents of the riparian zone made it safely to high ground? When I glanced up, the backside of a bear disappeared into the woods. Had the bear marveled at the flood?

Dead Duck Dam, one of the older and taller dams on the brook, had a great hole ripped through it. Water had risen in a great standing wave as it coursed through the gap at the peak of the flood. As the waters receded from Dead Duck Pond, the fine materials suspended in the water settled out and will provide rich nutrients for the meadow vegetation that will grow and feed hungry bears and beavers next spring. There I saw my first sign of beavers—a set of tracks led from sorry remains of the pond up to a fresh scent mound on the bank.

Upstream, at the pond where the beavers overwintered, the water level had dropped a foot but the dam held. The lodge also appeared to be intact. At the far side of the pond a beaver was busy with dam repairs. When she noticed me, Willow swam over and climbed up the muddy bank for a snack.

After a short visit, I continued my survey. At their new pond, the dams were indeed gone completely. The entrance to their lodge was now above water and a hole had been ripped in the side. Although I worried about the beavers, they would be all right if they managed to keep away from the debris-laden current. They had so many bank lodges; surely one or two remained above water. Even without shelter, the beavers would have been fine waiting out the flood. As for the destruction of their dams, each autumn, these beavers typically construct a brand new pond, lodge and larder. They would have plenty of time to rebuild.

Between their winter pond and their new pond is a third. Half grown in with emergent grasses and sedges, it is a bright green place with a big piece of sky above and a view of the mountain on the far side. Here I found three more beavers. When I sat down on the bank, two of them swam over eagerly, the two young beavers, Snowberry and Dewberry. I handed out apples and they settled down, making their proprietary squeak-whines. I then directed my attention to the third beaver, the one that approached uncertainly, the one with the very small tail, a new baby beaver! She swam up and prodded her siblings. They squeaked at her. She paddled over to the dam, ducked her head under the water, and came up with a pile of mud on her nose. She poked it onto the dam with all of the gravity and industry of her clan.

The only beaver missing was Bunchberry. For the past month he had been recovering from his leg wound, a process complicated by the development of an abscess. He could well be off surveying the damage or scouting for new dam sites. Still, even a beaver might have been seriously injured in that epic high water.

Night settled upon the pond with an intense blackness, and the universe sparkled above. I turned off my light and settled back to enjoy the perspective gained by a tour of deep space. In an infinity of blazing stars and black holes the events of this little planet seem safe and predictable, even with the odd tropical storm. I returned to earth when I heard the hum of a rodent greeting. When I turned my flashlight on, I found a large damp beaver sitting beside me hoping I'd brought him some rodent nuggets. Bunchberry had weathered the tropical storm, too.

The Balsams

$\sim$

THE BEAVERS' POND SITE that season provided even fewer amenities than Lake Dismal. They began construction where their stream ran through a scraggly second-growth forest of spruce and fir. The only place to sit was a tiny spot at the base of a fir tree and I had to keep my knees tucked up to keep my toes dry. There was not sufficient room for all five beavers to join me, so they had to come ashore in shifts to picnic. No vista of heavens or hillside redeemed this humble patch of shoreline; only a narrow band of sky could be seen, broken by the spindly spires of the conifers on the far side of the brook. To reach the beavers' new lodge, built against the bank just downstream, I had to pick my way through a miniature forest of balsam fir seedlings that concealed pitted terrain and a treacherous maze of toppled spruce poles.

Uninspiring as the location was, it needed a name so I could distinguish it in my journal. I decided it should share the name of a grand old mountain hotel where I once attended a conference. Located in northern New Hampshire, The Balsams stood regally, ringed by mountains, its stone facade duplicated in the still waters of its own small lake.

Tropical storm Irene did nothing to improve the appearance of the beavers' humble pond, for she left behind untidy piles of battered debris from upstream. Undeterred, the beavers stoically resurrected The Balsams. Long before the official arrival of winter, the beavers had gathered the enormous cache of branches that would feed the family through the long cold season. Their lodge, freshly plastered with a thick layer of mud, was reflected in the still waters of its own small (very small) lake.

Throughout December the pond ice spread and retreated again, but never fully sealed the beavers into their dark aquatic world. On January 2, after more warm weather and rain, I headed to The Balsams at dusk. Heat from the day radiated back into the winter-clear heavens and temperatures were heading down into the single digits. When I arrived at my usual seat I found that only the center of the pond was open. Farther upstream I located a place where open water reached almost to the bank, and the soggy shore had frozen enough that I could sit without getting wet. I called to the beavers and soon two appeared, paddling eagerly upstream. After some puzzling, Dewberry and Snowberry—the yearling and two-year-old—found their way over for their apples. Dewberry has grown to be almost as large as her older sibling, but I could still distinguish the two by behavior. Snowberry approached my boot and waited. Dewberry squeaked excitedly when she swam up, and then marched right up and climbed onto my lap in her search for treats. This little beaver has a special place in my affections.

Watching these two munching apples side-by-side, I pondered what the new year held for them. Snowberry, at two years, should have headed off to start her own family last spring according to the typical beaver schedule. Next spring both of these young beavers will likely head out.

As daylight faded, the half full January moon grew in radiance as only winter moons can, turning the ice to silver and casting spruce shadows across the pale Lilliputian forest of firs. As the cold began to seep through my layers and the apple bag emptied, I bid the Berries a happy winter and headed down to see if other beavers stirred.

From the center of the pond I could hear ice breaking and could just make out the shape of one of the big beavers clambering onto its surface for a grooming session. I crouched on the lodge, put my ear to its surface and heard the staccato chewing of a beaver dining. I imagined Sundew, the kit, snug in the warm but moonless dark of the lodge, resting on her elbows on her bed of dried sedges, devouring a twig she had retrieved from the larder.

The chill that settled that night and deepened over the next one allowed the ice to claim the pond. Safe from predators, well provisioned and warm, the beavers of The Balsams now had a season of leisure before them.

Ducky Diary

DUCKY AND GROWLER, once again in a head-waters pond, suffered no ill consequences from the ravages of Irene. We rejoin them in late November after several months of trials and triumphs. I rely on my beaver journal to recreate the sequence of events. Its smudged pages record not just the activities of the beavers, but which birds are singing, which plants are blooming, and other events of the day. These brief entries have the curious power to evoke the sights, sounds, and smells of each visit and the ambiance of the season.

On April 18, I noted "I arrive in light rain, sunset. A robin sings with gusto." I reported on the beavers in a similar abbreviated style. When I scan these words, the pond reappears before me in its April colors—the warm hues of sedges and leaf litter flattened by snow—and I hear the robin's irrepressible evening song. The journal reminds me that I conducted a sapling taste test that evening, allowing Ducky to choose between beech and striped maple. Ducky took a nibble of the striped maple, and then excised a four-foot section of the beech and began to eat the bark. In my mind I hear the rapid chopping of her teeth. Growler, her still-wary mate, also appears in my memory and scrapes together a scent-mound on the far shore.

My journal returns me to the pond on May 1, where I find false hellebore springing up along the streams and beaked hazel and dwarf ginseng blooming in the woods. From my couch in late November, I smell the warming earth and hear the whistling wings of a pair of wood ducks rising from the pond. As the light fades the first peepers begin their hopeful calls. Growler finally acknowledges me, and I almost hear his "hummy-growly" noise as he joins Ducky and eats striped maple.

On May 28, the journal notes, I arrived in the morning, not expecting to see the nocturnal beavers. In warm, bright summer light, I settle once again on the shore across from the lodge. As I do, a beaver exits from a den in the bank beneath me, speeds through the water, slaps its tail, and then turns and swims back into the bank tunnel. I wondered if this new den location was evidence that the pair now had kits. That morning I also recorded a synchronized dragonfly metamorphosis, the shrubs around me decorated with emptied larval exoskeletons and with dragonflies getting their bearings in their new winged bodies. I now hear again the soft short trills of the cedar waxwings that picked off the dragonflies as they took their first unsteady flights.

I did not make it back to check on Ducky until the middle of August. She, Growler, and any progeny, were gone.

I began my search for Ducky again and managed to locate three sites where beavers had recently settled. Among these was a pond on a remote hillside almost three miles, as the trout swims, from their former home. As I approached this pond for my first evening visit, I saw a beaver sitting by the shore. It swam off toward the middle of the pond, but did not give a warning tail slap. When I sat down and unpacked the apples, the beaver swam over. I had found Ducky. A second beaver continued his business, unconcerned—Growler. I saw no kits that night, nor any on subsequent visit.

To reach their new pond, these beavers would have passed through the territory occupied by Ducky's parents, and thus were the likely catalysts of events logged in the journal in mid-July. On November 28, I made a last visit to Ducky for the season. My journal entry notes that it was rainy and warm. Two peepers called as I approached the pond. Ducky ate apples and towed a sapling to the food cache outside their lodge. No kits were seen.

When the thaw arrives next March, and I pick up my journal and read that last entry, I will return to November and a night made soft and close by the warm rain and mist. I will see again the dazzling jeweled beaver, for the countless tiny droplets beading Ducky's dense, fine inner coat, sparkled in the beam of my headlamp.

Once again Ducky will launch herself into the deeper water, paddle out to the floating sapling, and position herself to seize the butt end between her teeth. With a few forceful tugs, Ducky and the tree will achieve momentum and move toward the lodge. I will not see Ducky as she works her sapling into the big pile of branches, but I will hear the splashing of her efforts. As the journal entry ends, I will hear the squeak-whine beavers make when they talk to each other—Ducky and Growler by their new lodge, ready for another winter.

Hideaway Pond

~

ONE OF THE PRETTIEST BEAVER PONDS I know occupies a tiny basin on the side of a mountain. It reflects a natural rock garden of ledges and trees on one shore. From the opposite shore the mountain drops away at your feet to another basin one hundred feet below, where another beaver pond reflects the sky. The rill that feeds the pond splashes in over a miniature waterfall in the shade of hemlocks, and departs in a long steep cascade to the lower pond. Small wonder that the site once inspired humans to bushwhack in a couple of miles lugging the materials to create a camp: metal roofing, a wood stove, windows, lumber... and finally the items that made it homey: curtains, bedding, books, cooking utensils... the message painted on an old saw blade stuck onto a stump invites visitors to *Enjoy the Stillness and Peace of Nature*.

While I admire the spirit of the endeavor, their sign's message is now ironical: their camp resembles a dump of sodden fiberglass insulation, odd pieces of metal, the collapsed remains of aluminum window frames, and numerous treasures turned to trash.

When I first found this pond several years ago, a single beaver was in residence. The peeper chorus was in full throat, and I sat by the lower pond to watch the beaver, my back to the cliffy knoll the derelict camp sprawled upon. No evening can be serene amidst the din of peepers, and if the noise weren't enough, several peepers pelted me in the back as they lost control on their descent to the pond. That evening was the evening that I watched a northern water shrew scurry across the surface of the water, the only mortal mammal I know of capable of such a feat. Heading home I practically tripped over a porcupine that had ambled into the meadow around the pond.

This spring, Ducky and Growler built a remarkable set of mud levees. These turned the meadow below their pond into shallow paddies where they could graze in safety and hoard the small trickle of water that flowed through their pond. I wondered how they fared as summer's drought intensified, so I headed up one evening to check on them. Indeed, although their pond still held water, their new pools were dry and I saw no signs of activity. I had a pretty good idea where they would be.

Peach faded from the sunset clouds as I arrived at Hideaway Pond and settled near the dam, the dam freshly fortified by armfuls of mud scooped from the pond bottom and patted into place by miniature five-fingered hands. A partially eaten harvest of twigs and leaves floated nearby, and so did one of the pond's occupants; Ducky swam over to the dam for an apple and floated in the water eating it. Growler swam over, too, and began eating the apple slices I tossed to him.

When Ducky had her fill of apples, she swam over to the lodge and climbed onto it to oil her coat. From beside her I heard a splash and then a third beaver appeared in the middle of the pond, a very small beaver. The kit paddled toward Growler, who floated near me munching an apple. I chatted softly to avoid startling the new addition to the House of Willow. Growler took his next apple and paddled away. The youngster and I studied each other until she noticed that her parents had departed. Alarmed, she slapped her tail and swam straight into the lodge.

Soon the kit ducked back into the water and began paddling busy circles in the middle of the pond. Growler swam to her and they greeted each other with beaver squeaks and swam off together to graze, an interaction more typical of mother/kit than father/kit relations. I might need to switch the sexes of the parents.

The moon disappeared behind the western hills. Hermit thrushes sang, creating surprising and beautiful chords, like Mongolian throat singers. The barred owls warmed up for a raucous night with a few shrieks and cackles. I watched Ducky work on the lodge and enjoyed the contemplation of a circle completed—a kit born, grown, and now with a family and pond of her (his?) own.

When the beavers leave this basin, as beavers have done countless times over countless centuries, the earth will quickly reclaim their lodge. The water level will drop, and wetland and meadow plants will grow in the nutrient-rich sediments. Deer, moose, bear, and porcupine will come to the meadow to browse. When beavers rebuild, a circle will end as a new one begins. Such cycles, small and large, have made life possible on this piece of rock for millennia.

It will take many, many years for the hideaway camp to disappear, and when it does, very little of it will enrich the forest ecosystem. It reminds me that we must think of endings when we make beginnings, for all things end. In nature, endings feed beginnings. By this, law charmed places exist—places where tiny frogs rain, mammals run across the water, and hermit thrushes sing.

Farewell

~

LET US PAY A FINAL VISIT to the beavers of
Popple's Pond. It is late July and they are enjoying
the ease of summer. Bunchberry inspects the dam and
then swims up toward the meadow to the north. He
seems to have adopted the habits of Popple and seldom
joins the picnic. Two year-old Sundew sits beside me,
reverently munching rodent nuggets. Her soft, reddish
coat has a rosy cast in the evening light. Balsam, the,
yearling, is in the lodge supervising the kits. If this year
is like the five that have preceded it, there will be two of
them. If this year is like the preceding five, only one will
survive the summer. When Willow hauls herself eagerly

ashore and clambers onto my pack looking for nuggets, I see that the eye that looked damaged on an earlier visit is now opaque and blind. While she has put on some weight this summer, I still feel her spine and ribs when I run my fingers through her coat.

Ducky and Growler are spending a second summer in the vicinity of Hideaway Pond, but upstream in a beautiful pond that they have refurbished. The gray treefrogs have finally retired to the woods once again, to be replaced by the glunking green frogs. The kit the beaver pair raised last summer, Fern, had been spending solo time in a neighboring pond, but with the arrival of siblings, she has returned to the main pond and babysitting duty. Growler recently stood up in the water near me, sniffing to figure out where the apples were, and I was able to confirm my suspicion. He is a she, and the mother of the new kits.

Funny little Snowberry, who waited by my feet (or a nearby stump) for apple deliveries, finally left her family this spring, sometime around her fourth birthday. Dewberry the Bold, my favorite, left the spring before, when she turned two. I have not located either of them, though I keep looking.

Mike returned to the confluence pond with his mate, and in that exceptional habitat they have managed to avoid trouble with humans for the past two years.

My other study beavers, I call them the Einsteins, were evicted from the confluence pond by Mike and his mate and spent the winter in a surprisingly small new pond in the woods. Last fall, they were hard at work preparing for winter in a somewhat more promising site, but when I checked on them mid-winter I found their pond abandoned. I found them in the spring at a different pond, though I saw only one of the adults and their kit from the previous year.

I think it noteworthy that neither Ducky and Growler nor the Einsteins were able to successfully raise a kit until their third summer together. Each pair then raised one kit. Beavers, at least in habitat like that found here, can hardly be said to be prolific.

Our culture has a shared image of the eastern forest primeval—a realm of trees supposed to be so dense that a squirrel could run from the Atlantic to the Mississippi without touching the ground. This image has fueled a conservation conundrum. How could such affiliates of meadows and shrublands, species like eastern racers, woodcocks, bluebirds, goldenrods, and asters, have survived in the deep, dark woods? If these species arrived in Vermont when land was cleared for agriculture, do we need to be concerned if these same species disappear in its absence?

We now know that such an image of the virgin forest is simplified. The concept of a steady state climax ecosystem has fallen out of favor. Disturbances like disease, wind, and fire create openings and are among the many variables that alter the course of succession. Furthermore, in regions of poorly drained plateaus, like Vermont's Northeast Kingdom and parts of the Adirondacks, numerous lakes, ponds, bogs, and marshes maintain open land and varied habitats. We also know that Native Americans created agricultural openings, especially along floodplains.

In the hilly lands of southern Vermont, though, another agent of forest disturbance created habitat for open land species—beavers. In the years that I have followed these beavers, I have had ample time to reflect upon the role of beavers as ecological agents. Books, articles, and the ecologists of my acquaintance agree that most, if not all, of the species that proliferated in Vermont's agricultural landscape once depended on beavers to generate their habitat.

To imagine the scale of beaver impact on the ancient forest, we must imagine how many beavers once inhabited North America. Research places the population somewhere between 60 and 400 million. If the upper estimate is right, there were once more beavers on this continent than the current population of humans in the United States (about three hundred million). We can add the 33 million Canadians, and might still have fewer *hominids* than there were beavers.

Beavers are country folk. Their population peak did

not result in concentrated beaver settlements. They were dispersed along the continent's waterways, with only one nuclear family occupying each pond or section of river. Large lakes might house a number of beaver families, but the density remained low. If we imagine the humans of our continent scattered in a similar pattern, we can begin to imagine the significance of the impact of these creatures.

Beavers and humans share a number of significant traits. Among them: both are large, palatable mammals that lack physical defenses; both alter the environment to make up for such shortcomings; both move into an area, rearrange things to suit their needs, and stay as long as the resources that attracted them last. Here is a difference: the alterations made by beavers greatly increase the habitat diversity, species diversity and ecological vitality of areas they settle in.

As long as the beavers maintain a pond, silt carried by the stream drops out in the still water. This results in cleaner water downstream and builds up fertile soil behind the dam. Further, sunlight penetrates and warms the still, shallow water in the ponds and creates optimal growing conditions for phytoplankton, which in turn feeds zooplankton, and on up the food chain to the trout and mink. Productivity of the stream system is enhanced.

The parade of wetland succession begins as soon as the pond has been abandoned. While the wetland flora species remain largely the same in each stage, dominance shifts as aquatic communities sprout more emergent plants, which then yield to sedge meadows and then to shrublands. This process is influenced by such variables as seasonal flooding, local hydrology, and the length and number of times the site has been occupied by beavers before.

In areas where beavers have been long established, they build relatively few ponds in new places, cycling instead through the areas they occupied formerly. In many sites, each occupation builds a more stable wetland environment with the accumulation of peaty soils. These spongy, organic soils hold moisture and support wetland shrubs and plants while discouraging plants and trees that prefer well-drained mineral soils.

The first beavers made their return to my watershed about fifty years ago. I suspect they found no remnant of the wetlands established by the beavers of yore, for the last were trapped and turned into hats about 200 years earlier. These pioneers moved into a valley of second growth forest. Stone walls, foundations, and old apple trees marked the places where human settlers put the beaver-generated organic soils to work.

After a half century the beavers have made some headway on the wetland restoration project. In places that have been dammed multiple times, the transition to semi-permanent wetlands is underway with denser carpets of sphagnum moss forming, and supporting such wetland plants as blue flag iris and round-leaved sundew. For the mere six years I have watched this place, I have seen alder swamps grow, forests flood, and meadows replace ponds as water recedes. So long as this tribe of short-legged wetlands restoration specialists remains in this wild valley, this natural endowment will grow.

I plan to stick around to see what happens for as many seasons as I can. In my ideal future, I will find myself, two decades hence, sitting on the shores of Popple's Pond. In this big wild place, I trust that there will still be beavers, and where there are beavers, every evening is eventful.

PART II
OTHER SKETCHES

Bear Diary

SNOW HAS MANY ALLURING QUALITIES, but if you fall under her spell you should know that she keeps no secrets. With each storm she spreads a blank page over the earth, and all who dally with her leave a record that anyone might see. Even wind and sun record their passage on her surface. The stories are prosaic, blunt, and honest. She is not one to embellish, nor to leave out personal details. The tales are not the less interesting for their lack of art.

If you wish to have secret dealings with snow, you may take heart in the ephemeral nature of her whisperings. If like me, you are fascinated by the accounts of those who have lived with her, you will need to go out often, for each day the old scripts fade and fresh ones appear. Some of the stories are so good that I feel compelled to preserve them in my own more enduring journal. The log entries of *Little Bear, November 2005,* are among them. Her story reminds me that the same candor must be applied to the reading of snow tales as was used in their writing.

∼

November 18—To one who thrives on snow, this winter had an auspicious beginning, with two good snow storms in October. Today, it seemed winter had settled in for the season. The snowy scene from my window should have been torment as I sat before the computer, but I was writing of a recent trip to New Zealand, and so absorbing were the rambles of my mind that I felt little inclination to leave my task even as day became night. For many hours my imagination wandered though ancient podocarp forests and clambered over alpine tussock grass with as strange a bunch of birds as have ever lived—poipoi, huia, moa, kakapo—and, indeed, many of these birds are now extinct. Not even the richness of this strange world could keep me indoors, however, when the moon spread its glow across the snowy field outside.

I set out at around 8 P.M. for what I expected to be a short, icy ski; but, oh, the skiing was divine. On top of several inches of firm base, a thin layer of powder danced with the edge of the skis, pushing them neatly into turns as if anticipating my wishes. As a fleet of clouds chugged across the sky, the nearly full moon teased from behind their cover, setting the ice-glazed trees asparkle and just as quickly muting them. Before I knew how it came to be, I reached the bottom of the hill nearly a mile from home. The wild music of a late flock of geese added to the drama of the forest. Why turn back? With skis in hand, I sought a place to cross the storm-swelled Hinesburg Brook.

As I scanned the bank, the moon pointed out tracks that emerged from the brook. I first thought *otter,* then *very large otter,* and finally *very small bear.* The snow and moon conspired to tell a tale and I intended to read it. Abandoning my skis, I set out to follow the peregrinations of Little Bear.

I followed the trail back up the hillside. The detail of the little prints in the fine snow was lovely, and a testament to their age; Little Bear passed this way sometime during the day. Here she scrambled atop a fallen beech branch and walked its length, leaving petite claw marks on the glazed surface. I noticed that her course was set to move from the base of one large tree to the next.

Gradually, I became aware that the mood of the night had shifted. Questions and their answers played tag in my mind. *Why is Little Bear alone?* Something has happened to Mama Bear. *What if Mama Bear is near and finds me following her cub?* Hmmm…I'll pretend I am looking

159

for something else. *Can little bears survive alone in their first winter?* No. Yes. No. *Perhaps these tracks will lead to the base of a tree where she is resting. Is it right to risk frightening Little Bear to gratify my desire to see her?* No. Yes. No. Well, I probably won't see her anyway. *Why is Little Bear alone?* Something has happened to Mama Bear.

~

I first met *my* bears in July while I was running along a trail in the spruce forest on the ridge above Hinesburg Brook. Mystifying grumblings alerted me to animal activity just ahead. Climbing atop a log for a better view, I saw the rear end of a bear as she loped into a glade twenty feet away. Scratching sounds from a nearby hemlock tree marked the hasty ascent of Little Bear. As I pondered the wisdom of taking a closer look, Mother Bear altered her course and pointed herself back at me. I made supplicatory gestures and turned away.

Since then, in my wanderings, I have learned that we share territory in the Hinesburg Brook watershed, the bears and I. I have found two beech groves that have helped them fatten. In each grove, small claws marked one tree, large claws marked its neighbor. Beneath this arboreal larder the litter of the feast was strewn—broken branches and the empty hulls of beechnuts. A vase of these beech branches, empty nut hulls attached, occupies a rather large space in my kitchen—a bouquet picked by the bears.

∾

Although Little Bear's wandering seemed without purpose, her pace was steady. She did not pause to look for food or evaluate her surroundings. Eventually her tracks led back to the brook, where I learned of a bear's superior facility in brook crossings. With little grace and high risk of a dunking, I followed her route across small debris dams. In one of the beech groves the tracks of many creatures pocked the snow beneath the trees: deer, turkeys, squirrels, and porcupines. There were the day-old tracks of a big bear, too, the claws of five toes registering in the broad print—Mother Bear? Little Bear did not slow. Above the grove she led me to the trail I hoped I would not see—the day-old tracks of a man. Little Bear followed these tracks for a short distance before turning and setting off down the ridge.

I had been out for a few hours and I did not recognize the terrain. I decided that within the hour I would turn back; it was past my bedtime, after all. Soon, however, Little Bear's tracks crossed a familiar trail and I knew where I was. Little Bear was leading me right back to my skis. Somewhere in the confusion of tracks in the beech grove I must have picked an older segment of her trail. "Good work, Little Bear," I thought, wryly amused. I was wrong.

As I started down toward the brook, the cub's tracks again met with the tracks of the big bear. Here Little Bear's trail changed abruptly. Back and forth over the big tracks she paced, sometimes stopping to scratch at them. At the base of a stump she spent some time resting and foraging.

The blood in the mother's tracks showed as a pink glaze. I brushed away the snow from the preceding night just to be sure. The tracks of the man were here, too.

Little Bear's tracks had become such a confusion that it took deciphering to learn that she moved off to the south, not toward my skis. I decided to let her be.

The tracks of Mother Bear and the man did lead to my skis. Both crossed the agitated waters on a high fallen spruce. I chose a lower crossing and began the ski home. I recognized the tracks I had skimmed by on my descent—Mother Bear and man heading northeast up the mountain. She had been moving fast, he more slowly.

The icy forest seemed the proper cage to contain my despair as I climbed toward home. Again, my mind parried questions: How it is possible for two creatures, the man and me, with nearly identical DNA, to have such different ways of feeling about bears? I imagined him following her, full of masculine pride. I imagined her forever leaping from his living room wall, her countenance frozen in a snarl. I wanted him to know of the travails of Little Bear. I wanted him to be sorry. Would I feel angry if a different predator had killed this bear? I would not be angry, but I would be just as sad.

I reached home and bed at midnight. For the rest of the night, in what must have been sleep, I followed the tracks of Little Bear. Through the trees I saw the ethereal forms of the vanished birds of New Zealand—poipoi, huia, moa. The wintery forest had become the lifeless haunt of doomed creatures.

~

November 20 —This morning the sun began to free the trees of their frozen burden. I set out into the icy cascade to look for Little Bear tracks again. My pack contained a five-pound bag of dog food. I planned to follow the big bear's tracks to see if Little Bear might linger where her mother had last been.

I picked up the bloodstained trail and in a dark mood I followed the tracks up the steep hill to the northeast. Halfway up, a tangle of young beeches forced me to take off my skis. I clambered through and picked up the tracks again.

In the stillness of the forest, the man's voice seemed surreal; "Do you know where you are?" I had never encountered another person in these woods before. Ten paces away he leaned against a beech tree, dressed in camouflage, his rifle relaxed. The expression taking cover behind the graying beard and brows was not affable. "Here is the man who killed my bear," I thought. My expression was not affable either.

I imagined him thinking, "Here's some nature nut trying to interfere with my hunting." When he began to give me terse directions to the nearest road I realized he did think I was lost. I assured him that I knew exactly where I was. We exchanged further scrutiny. He said, "Aren't you the woman who works for the nature center? I've been to a couple of their programs. I've seen you on TV. I recognize your ponytail, of course."

I wasn't ready for conversation. I put my question to him: "Can you tell me what happened to this bear?"

He looked away and began: "I can. A kid shot that bear, he was using a 30 gauge." He paused and looked back at me. "Do you know anything about firearms? Well, it wasn't a big enough gun to shoot a bear that size. She was a big bear, maybe almost 300 pounds. I met up with the kid that day half a mile back…" He gestured across the valley. "The kid was really upset and he told me what happened. I read him the riot act, in as nice a way as I could. I told him to get a decent gun, and don't shoot bears, stick to whitetails and grouse."

He continued, "I saw one of the cubs just up here in a tree that morning. It was a big, beautiful cub. There are two of them, you know. A buddy of mine hunts on the other side of this ridge. He saw another cub just before I saw this one. That sow is still alive." *That sow is still alive.*

The tension between us had dissipated. I moved closer to his tree. He said his name was John and he told me where he worked. He told me the young man that shot the bear was a good boy, just young and inexperienced. He told me the boy was very sorry about what he had done, both for botching the shot and for having fired in the first place. John thought he might know of this boy's family. They were a troubled bunch, but he had a feeling that this young man might turn out all right. I was beginning to be receptive to that possibility myself.

John told me that he and the boy followed the bear's tracks together. It was dusk by the time they came to the road and the bear was still moving fast. He arranged for a friend of his to help the boy track the bear the next day. The two of them set out in the morning and tracked the bear, "Must've been ten or twelve miles." She crossed Ames Hill, circled below the orchard on Barrow's Road, and headed through the woods to the far end of South Pond. By noon, the tracker told his exhausted charge that the bear had stopped bleeding and seemed to be stronger. He thought she'd probably survive. She had made her way back around toward where she'd left the cubs.

John and I spoke of other things then. We talked about the big male fisher that he'd seen in this valley. I told him about tracking Little Bear. He asked if I ever used masking scents to try to get close to animals and recommended a source. He talked, with a twinkle in his eye, about how he works with "a nest of liberals, but they're well-intentioned." They berate him for hunting and he chides their hypocrisy, those fastidious carnivores who buy packaged meat. "I don't know how you feel about hunting...?" he offered. I assured him, with only a little disingenuousness, that I don't have a problem with it. Not an ethical problem anyway. Not if one is hunting for food.

I caught myself apologizing, for the first time in my life, for scaring away the deer. What I meant to apologize for was annoying him by scaring away the deer. He reckoned it didn't matter. He wasn't really out there to hunt so much as to get outside, he'd use any excuse to get out of work to spend time in the woods.

"Did I see you had skis?" he asked. I explained that I seem to have some sort of pathological obsession with scrabbling through all kinds of snowy woods on skis. "Oh, no," he said, "That's good. I never knew anyone who skied off trails."

Looking out across the hillside that morning, we were comrades in our connection to the woods and our admiration for its wild inhabitants. Before we parted we plotted our routes. I had been spared a long and fruitless trek. I would head toward South Pond. He would move down the brook. Our paths would not cross again that day. As I skied off, I found myself hoping that they would cross again, though, someday.

Even the saplings that tripped me could not diminish my pleasure as I worked my way back down the hill. Mother Bear might yet live, and there were two Little Bears! Shedding my anger toward the bear hunter was no small portion of my mood alteration. John had reminded me of the disservice imagination can play in the judgement of others. I pondered how frequently stereotypes inflame conflict among potential allies, especially those who share a need for forests. I promised myself that I would approach future encounters with a more open mind—who knows what form a watcher of the woods might take.

~

The best story I have ever read was written in the script of the winter forest, by feet in the snow. When I found the tracks by the marsh near South Pond that day, the prints had begun to soften. Still, they had been there, anyone might see it, Mother Bear and Little Bears together.

Porcupine Paths

~

MOST SNOW STORIES are written by the ener-
getic creatures we expect to be active all winter,
like foxes, weasels, snowshoe hares, and squirrels. We
expect the slower animals—woodchuck, bear, skunk—
to spend most of the winter sleeping. There is one creature that refuses to conform to this stereotype—the
corpulent quill pig waddles blithely through whatever
winter dishes up.

To find a porcupine trail, strap on your snowshoes
and head to almost any location that combines rocky
terrain and hemlock trees. These elements rarely occur
together without at least one resident porcupine. Look
for a trough about ten inches wide that has been traveled
repeatedly. By following the trail in one direction you will
find the tree (or trees) where the porcupine feeds. In the
other direction you will find the porcupine's sleeping
quarters.

Porcupines subsist on a winter food that is high in
quantity but low in quality—the inner bark and small

twigs of trees. I see them most often in hemlocks, though each porcupine includes one or two additional tree species in their winter diet. This high fiber diet is slow to digest so there is a limit to the number of calories a porcupine can acquire. Surviving winter is a matter of careful economy. In sub-zero temperatures, porcupines must increase their metabolism to stay warm. Their success depends upon fat accumulated in the summer and fall, and the weather.

When I follow porcupine trails to their dens, I typically find they've selected the base of a hollow tree, a crevice in a ledge, or a rocky shelter in the talus at the base of a ledge. A rocky nook might seem like a frigid place for a winter nap, but if you were to remove the quills from a porcupine you would find a very dense undercoat and long snow- and rain-repellent guard hairs. Very little heat is lost through a porcupine's coat. The quills insulate, too.

Even uninhabited dens can be recognized readily. Porcupines are one of the few creatures with indoor bathrooms, though they lack the benefit of plumbing. While a mountain of droppings might look messy, by the time a porcupine has finished digesting, the remains are compact crescents of sawdust and might be one of nature's better beddings. I suspect that heat may be generated by the composting process, as well.

I'll admit that the snow diary entries of a porcupine usually describe a life of tedious routine, but winter is the slow season for porcupines. During the summer and fall the sphere of porcupine life expands.

In the spring female porcupines give birth to a single porcupette. I once had a chance to interact with a baby porcupine while visiting a wildlife vet. While too prickly to pet, Quilly could compete with any wild baby in scampering, hopping in the air, and tail chasing competitions.

More recently, a mother and baby porcupine summered in a basswood tree near my house. They often fed on low branches within easy viewing range. I spent more time than I care to admit lounging beneath that tree watching the porcupines' leisurely activities and listening to their porcupine talk.

The talk between mother and young is varied and mild, the conversations that take place between adults is anything but. One autumn night I was awakened by piercing shrieks. The source puzzled me only briefly since I was close enough to hear that each shriek was punctuated by the sliding whine that is characteristic of porcupine talk. Sleeping through the din was not an option, so I crawled out of my tent and went to see what the fuss was about. The two porcupines sat next to each other on a branch at about my eye-level. From their activities I judged that one felt amorous, while the other insisted that she was not in the mood.

I have since learned that this is typical of the porcupine mating season. Female porcupines send out olfactory signals that let males know they will be interested in mating, "oh, sometime in the next week." This gives the males time to assemble and demonstrate which of them is fittest to sire the next spring's baby. One fall, squeals, shrieks, and whines were a part of the nocturnal music for a couple of weeks, as a female porcupine and at least one or two very large suitors dined on apples in the tree outside my bedroom window.

Occasionally people call me to describe a scream they heard in the woods, imagining it must have been the mating call of a fisher or some poor creature meeting a terrible end. When I give an imitation of a porcupine shriek, many such sounds turn out to have been merely disgruntled porcupines.

~

One cold November night, heading home from the beaver pond, a strange sound could be heard part way up the hillside above me. When I paused to listen I recognized a porcupine whining with noteworthy petulance. Intrigued, I began a noisy ascent toward the sound. I soon found myself surrounded by jumbled boulders at the base of a ledge, fine porcupine real estate. The porcupine complaints continued and I saw the backside of their source, a substantial rodent, his waddling retreat punctuated by querulous mutterings. Such annoyance must have been provoked, and I shone my light about to find its source. Sure enough, when I looked into a cave formed by the boulders, an equally ample porcupine was in residence. I surmise that the noisy porcupine felt some proprietorship toward the commodious den before me, and had returned home to find a squatter. Who wouldn't complain?

And so into the November night went porcupine and I, and into the season of stillness and snow.

Subnivea

~

O NE BRILLIANT MOONLIT NIGHT, with temperatures dropping into the sub-zero°F range, I
headed to one of the wildest open places I could reach
without risk of frostbite. This large wetland meadow
near a hilltop provided the perfect stage for the moonlight show, so I sat down to admire the sparkles and
listen to the music of the winter night. The sparkles were
there all right, but I heard not a sound—not a breeze,
not the low rumble of distant traffic, not a plane, not a
woof, or a crackle, or a hoot. This crystalline night en-
forced the notion that winter is a lifeless season. I wasn't
fooled. I knew that within a couple of feet of where I sat
there existed a realm in which the inhabitants, oblivious
to the deep chill, busied themselves with the activities
of their lives. Each winter this realm is created anew, and
although it is as big as winter itself, it is a world we cannot enter, not even through a magic wardrobe. Scientists
call this the subnivean environment—the zone of life
beneath the snow. I call it Subnivea.

All winter the earth slowly releases heat, some stored over the long days of summer and some radiating from the planet's molten core. Snow, especially fluffy snow, is one of nature's fine insulators. Under more than about eight inches of snow, the surface of the earth remains within a few degrees of freezing regardless of the weather above. Moist warmer air moves up through the snowpack until it cools or escapes the snow. This process transforms the lower layers of snow into a weakened lacework of crystals called depth hoar. Within the depth hoar, caverns open and tunnels are excavated. Here the plants and animals that populate Subnivea spend the winter.

Many of the mammalian residents are voles, a tribe of peaceful vegetarians. They spend much of their time in furry piles in soft nests, occasionally heading out to maintain their tunnels and eat. The voles of my forest are redback voles. True to their names, the top of their backs has a wash of rusty color. Red-backed voles eat seeds, fruit, fungi, berries and leaves. They prepare for winter by creating stores of food. I suspect red-backed voles were responsible for a pile of cut sedges I found dragged beneath a rock on a wooded hillside last fall. One of the most charming things I have learned about these voles is that they sing, sometimes in groups. I love to imagine a group of them trilling in a Subnivean chamber like miniature coyotes. Meadow voles populate nearly every open area. They maintain neatly clipped tunnels, outhouses, and sleeping chambers beneath the grass thatch.

Elsewhere in the neighborhood, the shrews, solitary

flesh-eaters, manically seek prey. Shrews are renowned for their metabolic extravagance. They need a nearly constant supply of food. A shrew that hasn't dined in the past few hours is in on the brink of starvation. They emit echo-locating clicks to navigate. If their territorial borders are breached, shrieks and fisticuffs ensue until the intruder is driven away. While invertebrates have the most to fear from shrews, even the voles and mice must tremble when they encounter these hunters in the dark tunnels. Shrews use toxic saliva to paralyze their prey, and sometimes include other small mammals on their menu.

Each fall the shrews and voles probably join me in praying to our chosen deities for snow. During the fall these small, active mammals become especially vulnerable. Tiny bodies lose heat quickly, so finding food and maintaining a well insulated, dry nest becomes critical. Cold rains are not just a matter of discomfort to these little beings. Once secure in the shelter and warmth of

Subnivea, the mortality rate for small mammals drops. When snow is deep, Subnivea can be warm and secure enough that some creatures can keep raising families right through the winter. Meadow voles are especially prolific. Females can begin breeding within three weeks of birth, and can give birth to a new brood as often as every 22 days.

The patchy snow that we had after a January thaw provided a window into this world. In the fields I could see the slightly humped grasses that revealed meadow vole tunnels. When I lifted a section of icy roof, I could see the neatly clipped, winding corridors and the soft balls of frayed plant fibers that were their nests. I tried to see what the voles might eat. I found a variety of evergreen weeds and ferns, seeds, fragments of leaf, and discovered that the bases of some of the grasses were still green. In the woods I found maple seeds and nibbled hemlock needles among the leaf litter.

I found much less insect sign, but the hordes of insects that disappeared with the frost are not really gone. I trust the shrews know how to locate the eggs, larvae, pupae, and dormant adults ready to spring to life with the thaw. Some insects, spiders and other invertebrates remain active all winter in the shelter of Subnivea.

Even given mild temperatures and sufficient food, Subnivean life is not without peril. The little mammals in their wintery world are relatively safe from most predators when the snow is deep. In shallow snow, fox and coyote can locate warm bodies beneath the snow, and with a pounce and some quick excavation, Subnivea is violated.

There is one creature that moves with grace and comfort between Subnivea and our world, the tiny short-tailed weasel. This weasel, also called an ermine, meets many of its own winter energy needs on incursions beneath the snow. I have read that they use fur from their suppers to line their nests.

Subnivea can also become a hostile place during thaws and heavy rains, which can seal portions of Subnivea in impenetrable ice, or flood tunnels and nests. I noticed that there were many more tracks of mice, shrews, and voles on the surface of the snow after a heavy rain than I had seen earlier in the winter. It is likely that warmer winters will make life less secure for these denizens of the deep.

Spring magnifies these hazards as snow melts and Subnivean tunnels collapse. If small mammals enter the spring season in a weakened condition, they will be especially vulnerable to the cold and wet that must be survived before the months of warmth and plenty.

When the snow does disappear, you can explore the archeological remains of Subnivea. You will be able to find streets and tunnels, remnants of meals, food caches, and nests. The nests of meadow voles are easy to find—grapefruit-sized balls of grass with an inner lining of soft, shredded plant materials. Their tunnel system is well mown by little vole teeth. You might even find the chamber used as the communal outhouse. Leave things as you found them, though. While Subnivea may be gone, its inhabitants remain, and if they're very lucky they will populate the dark warm zone beneath future snows.

February Hare

∽

M Y ONLY COMPANION on many winter evenings is a lop-eared rabbit whose chief merit is extreme cuteness. Roo seems content just to sit, flop, or hop about in my proximity, thoroughly domesticated. He has no notion that not far beyond the walls of this warm illuminated structure, his cousins, wearing winter white, hop through a snowy glade in the moonlight.

I seldom see them, but there is no mistaking their presence. When I go to the snowy glade I find their tracks—those distinctively widespread toes of the snowshoe hare. One afternoon recently, I decided to see what the tracks would tell me about their activities. The tracks in the glade showed that the hare were feeding on birch and cherry twigs and the twig tips and needles of white pine. They were not heading down into the tents formed by small spruce trees, spots where I always imagined snowshoe hare lurking. Instead they seemed to accumulate on small rises where hares could see in all directions.

The glade was full of tracks, but they did not go south into the field or to the west where the spruces grew tall. They did head into the dense young forest of spruce and fir to the north, however, and so did I. When I started out, I was simply following tracks through a familiar woods. Somewhere along the way, however, I realized that I was completely disoriented. I had entered a low, close, dark world—the bustling metropolis of the snowshoe hare. I began to think, that, like Alice, I had followed the white rabbit down a hole and into Wonderland. The tracks no longer seemed random and confusing. Instead they demarcated a network of paths that led between areas of concentrated activity. Red-backed voles, a couple of flying squirrels, and a coyote also recorded their activities. The freshest tracks of all, almost as fresh as my own, were made by a fisher that loped down the rabbit hole just before I clambered out of it. Two hours had passed and darkness had fallen when I found myself back in the open and on a familiar trail just below my house.

Snowshoe hares are not, technically, rabbits. They are in a different family. Hares have longer legs and longer ears, as a rule, and their young are born fully furred, eyes open, and ready to hop. Rabbits give birth to hairless, helpless young. In our region, snowshoe hare have a strong affiliation with spruce-fir forests and need snow to thrive. Not only does snow give them an advantage over predators when they deploy their snowshoes to escape, but they depend on snow for camouflage. Their coats change color with the day length. A white hare in a year with no snow might as well be ringing the dinner bell for predators.

When I drive the ten miles to the lower elevation farm that houses the Bonnyvale Environmental Education Center, I do not find snowshoe hare tracks. A few times, though, there have been the tracks of a cottontail rabbit. The eastern cottontail has been in decline in our area, though there is some suggestion that the trend might now be reversing. I recently followed the tracks of one of these cottontails. When I first found the rabbit's tracks, they seldom ventured far from a brush pile. Within a few weeks I could follow its tracks thirty feet up a hill to a couple of apple trees and a thicket of grape vines. After following snowshoe hare tracks, I expected that the cottontail would also choose to rest in a place that offered good views and multiple escape routes. I was surprised to find many tracks leaving and entering a tunnel beneath a woodpile. No tracks indicated another exit. On a few occasions I have seen fox tracks pass through the small world of this cottontail, but so far the cottontail's survival strategies have succeeded.

Cottontails do not change color in the winter. They need to be in dense cover to hide. I have read that cottontails (and other rabbits) use *flash marking* to foil pursuit.Here's how it works: predators home in on the rabbit's tail—that flash of white with each bound. The rabbits, however, quickly switch direction and then freeze. The predator goes to the spot where the last flash was seen and is bewildered. My colleague, Deb Smith, watched this happen when a cat was stalking a cottontail in a meadow. From her vantage she could easily see where the rabbit had hunkered, but the cat just stood where it had last seen the rabbit's tail, clueless.

At certain times of year the stories in the snow might require some censorship. Foxes, bobcats, coyotes, raccoons, squirrels, and Roo's relations choose late winter for courtship. Hares and rabbits take reproduction very seriously. Not surprisingly their courtship ritual is among the most zealous. Perhaps you have heard the expression *mad as a March hare?* In our area, rabbits and hares often begin courting a bit before March. In mid February the boxing matches and posturing will commence as males establish their position in the hierarchy and females express their disdain for them all. Sometime in the following weeks, however, the females will change their minds. Here is how their courtship is described, though I don't know how closely a particular pair follows the script: The male will rush at the female, and if she is interested she will leap over him while urinating. The pair will continue this frolic, leaping over each other, urinating, chasing, and boxing, with occasional pauses to mate. I once found tracks in a little meadow that could only have been made by snowshoe hares indulging in such a bunny bacchanal.

Sometimes I think Roo has a pretty good life—fresh vegetables delivered year round, warm, dry accommodations, and hardly anyone tries to eat him. Still, if he knew about the white hares gamboling in the moonlight just beyond these walls, I wonder what he'd think of his bargain?

Mink Envy

~

RIVERS AND STREAMS ASSUME a surreal beauty in winter. Beneath the slabbed layers of ice and snow that accrete along the banks, the elements interact to create dazzling crystal sculptures. Skiing along such wild watercourses is one of life's great pleasures. Why, then, do I find myself returning from such excursions suffering from a terrible yearning? The affliction, mink envy, is not uncommon in humans, but I believe in my case it has assumed an unusual form. Yes, the fur of the mink is dense, dark, and exceedingly lustrous, especially in the early winter, but it is not the mink's coat that I covet, it is the little creature's winter haunts. How I would love to be a mink, at least for a little while.

Tracks provide a vicarious mink experience along nearly any stream in winter. Mink trails often begin or end at small tunnels or crevices, or at the water's edge. I can see something of the splendor of the caverns the mink explore, and suspect that if I could see these plac-

es from a mink's perspective, the world would be sub-
lime indeed. I see this world through rounded windows
in the pillowy snow; sometimes the windows are large
and reveal deep, inviting pools, with water rushing into
them over miniature waterfalls. Suspended from the ceiling above these cascades, tiers of icicles look like ornate
balconies in a fantasy castle.

Sometimes the snow windows are small, and provide
just a glimpse of ice and rushing water far below. In other places I can only hear the drama; where the water
drops in an enclosed chamber, it rumbles like thunder
beneath my skis.

Imagine the world of muted light and damp air with
the only sound the gurgles and splashes of rushing water.
The mink hops and slides through mazes of sparkling
stalactites, moving with ease between water and earth.
Researchers have found that a mink with a stream or
river as the core of its territory will claim about a mile to
a mile-and-a-half of the watercourse as its domain. Even
within the dynamic world of water and ice, each mink
must develop a familiarity with the splendid ephemeral
galleries and grottoes.

My image of the mink that make these tracks has
been shaped by a number of encounters. Once, on a ski
visit to a beaver pond, I saw a mink exploring the top of
the abandoned beaver lodge. The mink spotted me, too,
and after expressing some initial concern over my proximity (a distance of about fifty yards), chose to ignore

me as soon as I sat down and pulled out my binoculars. For several minutes he explored the snow caves created by sticks protruding from the lodge, then paused and scanned the open water below. In the next moment he bounded down the stick pile and with a graceful hop and insouciant flip of his tail dove into the opening in the ice. Bubbles and thrashing roiled the surface of the water for just a couple of seconds before the mink leapt onto the ice again, shook himself, and gave a self-satisfied hop. An eight-inch trout protruded magnificently from the little predator's jaws. The mink followed his lunch with a roll in the snow and then loped out of view behind a ridge in the ice.

I encountered another mink one morning as I ambled along the shores of another pond. This mink seemed to have detected my approach and was in the process of bounding off into the woods. As I have mentioned before, I am testing a theory that many animals recognize a friendly tone of voice, and assuming the speaker is outside of their mandatory flight zone, that they will interpret harmless intentions correctly. When I spoke in this voice to the mink, she paused, looked at me, and seemed to relax. She then turned and headed back toward the pond and began busily exploring the shoreline. After a minute or two, she sprang into the pond and sped through the shallow water with the same undulations that mark a mink's terrestrial travel. I like to think this encounter provided further evidence to support my theory, but must admit the possibility that the mink returned to the water's edge when she detected me because she felt more secure there.

I see otter tracks along winter rivers too, and while I suspect otters enjoy life even more than mink, they can't go all of the places mink can go. Otters are hefty beasts, more than forty inches long and weighing up to twenty pounds. A large mink is half that length, and weighs only two pounds.

Like otters, mink are not above sliding on their bellies as a means of moving in snow. I once had the good fortune to spend a winter day in the company of some of Vermont's finest naturalists. I didn't imagine that we would encounter anything to astound such a group in an ordinary snowy forest, but the mink slide we found did just that. A mink belly slide leaves behind a smooth shallow depression three to five inches wide. This particular slide looked like the descent of a one-legged skier; we could see a stretch of about 120 feet of uninterrupted belly slide. I had to clamber down to the end of the slide to confirm that a mink's belly was indeed responsible for the trail. There I found the paired bounding tracks made as the Olympic caliber belly-slider left the scene.

Recent storms brought a bounty of snow and buoyed my hopes for an extended season of skiing along wild rivers and streams. My joy would be boundless if I weren't so often reminded that another dimension exists, a wondrous world I cannot enter. Yes, I want to be a mink! Before you judge me mad, I encourage you to get out on a winter river and see if you are immune to mink envy. I'll wager you'll return with sympathy for my condition.

Offseason at the
Whispering Pines Diner

~

WHEN I MOVED TO THE HILLS of Marlboro,
I knew that my diner would have a different cli-
entele than the one I operated in the valley. This land of
conifers and cold is the undisputed domain of the red
squirrel, with chickadees, nuthatches, jays, and grosbeaks providing the local color.

I wish I could say I know the regulars at the Whispering Pines. I've been working to recognize the different
squirrels. I now notice variations in color and markings,
fuzziness of ears, carriage of tail, personality quirks and
facial profiles. Still, there are only a few I can greet with,
"And will you have the usual today?" Even if I manage
to master the squirrels, there will still be hordes of avian
look-alikes. I see why a good diner waitress calls everyone *dear*. I find, however, that if I make the rounds of
the tables offering cheerful flattery, the birds don't mind
what I call them, but continue chatting and eating.

Like any good greasy spoon, my diner provides the social hub for year-round residents once the summer folk have flown south. One of my customers has made me especially aware of the social value of diners, a hermit thrush that dropped in one very cold day in mid-December. The hermit thrush, Vermont's state bird, is esteemed by bird song connoisseurs worldwide. In the winter, hermit thrushes have the good sense to fly someplace warm. What was this bird doing in Marlboro?

While chickadees hopped about energetically, the thrush shivered and always kept one leg tucked into her belly feathers. I suspected the activity of the other birds drew her to the buffet, but she looked baffled by the offerings. She would cock her head and stare, and finally scurry to peck at a bug-like spot. I watched her through binoculars, but couldn't tell if she had selected something nourishing or just a fragment of sunflower seed shell. The cold day gave way to a much colder night, and I thought this hermit thrush would be eliminated from the gene pool. When I saw her the next day I was surprised, but not optimistic. When she returned for a fourth day and wasn't shivering, I began earnestly scheming about ways to make my menu appeal to a thrush.

I knew thrushes ate insects, and from her concentration I imagined that she waited for something to wriggle. The first menu offering I tried was suet chopped into bug-sized pieces. I watched, but she didn't seem to choose those morsels. I sprinkled them with thistle seeds to make them look more like insects. She was not impressed.

When the weather warmed enough to get snow fleas and other winter insects moving, the thrush disappeared. On cold days she'd be back at the Whispering Pines. As she began spending longer hours at the diner, I became concerned that she was filling up on specks of cellulose, though her survival suggested she was getting nourishment.

One of the people I call when I have bird questions suggested the diner menu include berries. I stocked up on currants, dried cranberries and apricots. I spiked them on the buds of the trees she perched on and scattered them in the places she foraged. I never saw her touch one. I chopped them into little pieces and sprinkled them with thistle seeds. She ignored them. Finally I tied a thread to a sprinkled currant and ran the thread through the window into the house. I waited for the thrush to hop within range so I could make the bug wiggle. She ate at other tables that day.

While I watched I noticed that she (finally) seemed to be making deliberate selections—light-colored items that had some substance. I inspected some of the chaff she hunted through. While sunflower shells made up the bulk of it, I did manage to find a few seeds and seed fragments mixed in. I decided to change the whole diner menu.

No more confusing shells; I'd feed sunflower seed hearts, with thistle seeds and bits of suet on the side. I also found a one-quart tub of dehydrated mealworms that looked tasty. The diners approved. The next day every plate was pecked clean. By the end of January, the hermit thrush looked like a competent winter resident with good feeder skills.

One weekend in February, I noticed that the diner was empty. When I went outside I heard the avian patrons gathered in the treetops. What noise! Grosbeaks chirped and preened. Nuthatches hopped up and down the trunks, honking like miniature bicycle horns. Chickadees added their chatter to the din. These birds were not mobbing a predator; their sounds and activities were relaxed. With days growing longer and food in their bellies, everyone had moved out to the street for an all-species celebration. Up on a branch in the thick of it I could see a tail flick up and descend. Sure enough, there was the musical call note of the lone hermit thrush—lone but not lonely—part of the chatter of the locals.

A Long Winter's Nap

FOR THE PAST TWO SUMMERS I have spent time as the personal servant to a couple of *semi-fossorial sciurids* (according to one scholarly source)—members of the squirrel family that spend some of their time underground. While my small striped masters offered no direct compensation, the hours I spent attending to their demands were among the brightest of those warm green seasons.

Jackson trained me first. He appeared on the stonewall by the lilac bush each morning. In no time I learned to bring him breakfast, and then lunch, and then… well, you know the rest. By the end of that summer many pounds of seeds and nuts had been transported to Jackson's larder.

This year Electra, a dainty chipmunk, took my level of training even further. Chipmunks, as it happens, lose their fear of people rather quickly if one is appropriately humble and offers food. It didn't take Electra long to show me where she liked to eat and what she liked to eat. I fancied that I trained her to take the next steps; I soon had her sitting next to me while she ate, and then from my hand, and then on my hand, and then on my lap. The other neighborhood chipmunks were quick studies and at least four would appear when I went outside. These chipmunks never failed to remind Electra that she was just a little no-account first-year chipmunk whenever they had the opportunity. Electra however, used her first-year cuteness to buy protection from the big provider of seeds. I left piles of seeds for all of the chipmunks, but Electra ate in safety on my raised hand. I still recall the surprisingly light touch of her tiny feet, and the warm fur of her belly as she delicately vacuumed the seeds from my hand (yes, her full name was Electra Lux). How her cheeks could expand! I once counted out a pile of seventy seeds for her. She sucked up every one and had room for more.

Naturally, it is difficult to study the winter activities of chipmunks. What little we know comes from a few studies of excavated burrow systems and from captive chipmunks. According to these studies, the burrow systems of chipmunks range from a simple one or two entrance tunnel with a single chamber to elaborate networks with many tunnels and chambers. One tunnel system was found to have thirty entrances, although not all were in use. Such systems typically have one nesting chamber lined with leaves or leaf fragments. The burrow will also have storage chambers. The tunnels are fairly shallow, with the deeper ones only thirty inches below the ground surface.

Chipmunks have long been snubbed as *not true hibernators*. It is now known that during winter torpor their respiration rate drops from 60 breaths per minute to 20, and their temperature drops from about 100°F to 42-45°F; this, some scientists argue, is slow enough and low enough to put them in that selective group of true hibernators. Unlike other members, however, eastern chipmunks do not retire bulging with fat. They must awaken every once in a while to eat. Just how often is uncertain, but some have been observed waking every six days in captivity.

As I watch the red squirrels eating outside my window, wind blowing wet snow into their fur, I wonder at how they contend with the elements during the long months of winter. It is more pleasant to contemplate Jackson, Electra, and their clan snuggled solitary in their soft nests in the ground. Every week or so they might yawn and stretch, and then wander down some dark corridor to a well-stocked pantry for a snack. Then it's back to bed to continue the long winter's nap. Knowing they are snug underground makes me feel less guilty about hoping the winter will be long and snowy. Whether it is or isn't, spring will surely come and one of its pleasures will be the return of the chipmunks. They are likely, of course, to make many requests and will probably find that I am easily trained.

Flying Circus

ONE NIGHT LAST FALL the neighbors' motion detection light went on. I knew they were away on vacation so thought I'd better investigate. A slight rustling in one of the apple trees betrayed the culprit; there, in my flashlight beam, a flying squirrel clasped an apple in a furry embrace. The petit squirrel was so engaged in its meal that it paid no attention to me and I admired it for several minutes.

Few can dispute the pleasures of such a task. The back of its thick lustrous fur cape had a border of black. The belly side was a snowy white. The paws gripped the apple like pale-gloved hands, and it watched the night through immense dark eyes. When I turned to see if it had company in the other apple trees, it disappeared. That is the way of flying squirrels.

I have had ample opportunity to observe flying squirrels in a less-than-wild situation. One winter I provided a temporary home to nineteen of them. They resided in a large outdoor enclosure equipped with branches, two cozy nest boxes, and all the nuts, seeds, and dried fruit their little hearts could desire. The situation was not ideal, but I judged it far better than their other option—a mid-winter release in an unknown forest with no cached food.

Mind you, the squirrels thought they already had a pretty nice situation. They had found a crack that allowed them to squeeze beneath a roof—a warm, dry, soft space to spend the winter. They were so happy there! Their nocturnal celebrations lasted much of the night, or so it seemed to the people trying to sleep beneath them. The squirrels were lucky that none of the flying squirrel recipes concocted during those sleepless nights were prepared. Instead, these people lured the squirrels into a live trap, one by one, and brought them to me.

Each night I would head to the Nut House at dusk with a jug of warm water, flying squirrel provisions, and my headlamp with the red filter. Some of the squirrels remained too shy to come out when I was there. Others perched like gargoyles, shoulders hunched, noses pointed down, on the roofs of the nest boxes, daring me to come closer. My favorite squirrels popped out eagerly and came to see what I brought. They would settle down on my hand or arm and tuck into supper. When they were finished the little squirrels would hoist their skirts and scamper out my arm, down my leg, up the walls, then leap onto my head. By this point the gargoyles would join the melee, and the shy squirrels would peek from the nest box openings.

Although they may well be more abundant than red or gray squirrels, flying squirrels seldom reveal themselves. When they land on a tree, they scoot to the far side of the trunk so quickly it can look as if they've vanished. Such evasive maneuvers help foil owls. Two species of flying squirrel can be found in Vermont—the northern flying squirrel and the southern flying squirrel. They are as good as indistinguishable unless you have them side-by-side to examine the roots of their belly fur. Northern flying squirrels average just under 11 inches, nose to tip of tail. Southern flying squirrels are smaller, about 9 inches. Chipmunks, for comparison, average about 9.5 inches long. The belly fur on northern flying squirrels is gray at the roots, while the belly fur on southerns is white to the base. The two do have different habits. The southerns eat a typical squirrel diet—nuts, fruits, insects, buds and other plant parts. Northern flyers will eat these things as well, but are fungus specialists. Fungi are a preferred food, especially those with subterranean fruiting bodies that resemble puffballs. Research has shown that flying squirrel scat provides an ideal growing medium for these fungi, so when the squirrels eat them, they transport and plant the spores. These fungi become part of the structure of tree roots and are important, sometimes even essential, for tree health, helping the roots to absorb nutrients and moisture.

When the frolics of the flying squirrels in the Nut House began to include mating, I did the calculations and determined the date when they'd need to be released. In early April I located some good habitat and put up nest boxes and feeding stations. One day, while the nineteen little squirrels enjoyed their cozy diurnal slumber, I covered the openings to their nest boxes and transported them. That night they would once again scamper to the treetops and launch themselves into the air. Their new home was about thirty miles from my house, as the squirrel flies, and it seemed unlikely that they would return to plague my patient neighbors. I was sorry, though, that my nights would no longer include visits with those beautiful gliders.

Of Squirrels and Oaks

THE SNOW HAD FINALLY STOPPED FALLING, but the air remained opaque with blowing crystals. In sheltered places it had settled to a foot of fluffy white. All morning I watched out the window to see how Priscilla would manage in this transformed world—Priscilla who had never imagined snow before—Priscilla whose legs are just three inches long.

I finally saw her descending the little pine at the edge of the woods. Six feet from the bottom she hesitated and then leapt. She landed in an explosion of snow and bounded intrepidly toward the house, a furry porpoise. Halfway to her destination she vanished. A disturbance in the surface marked her progress, and four feet farther, still traveling straight, she burst to the surface again. I trotted to the door to greet her and offer her breakfast.

Priscilla is the only one of my squirrel orphans who comes to visit me daily. I occasionally see some of the other squirrels I have raised and released, and know that at least a few have not just survived but reproduced.

These squirrels no longer depend upon me. They depend upon a more primal relationship, that between the squirrel and the mighty oak.

The squirrel offers the oak outward mobility. By transporting acorns from beneath the parent tree, squirrels increase the chances that a resulting seedling will have enough sunlight to grow, and, if it matures, that unrelated oaks will pollinate its flowers, conferring greater genetic fitness on future generations. Further, by hiding the acorns, the squirrels remove them from the sight of the many birds and mammals that would feast on them. Burying the acorns also increases the chances of germination—if the squirrel does not return to eat them, that is.

The squirrel's needs are quite different from those of the oak. The squirrel must gather and cache a sufficient store of nourishment to survive the lean months of winter. Acorns provide quality nutrition in durable packaging. The relationship between squirrels and oaks has been refined by the ancient dance of co-evolution, and is surprisingly complex. It becomes even more interesting when a third player is introduced, the human scientists who study squirrel caching behavior. My best source of this information, a 2001 Smithsonian publication, *North American Tree Squirrels,* was written by two of these scientists, Michael Steele and John Koprowski. I find myself nearly as fascinated by the research they contrived as by their insights into squirrel behavior and biology.

Among the questions they set out to answer was whether squirrels exercise discretion in deciding which nuts to hide and which to eat and, if so, which characteristics are important. North America hosts many species of oaks, which are divided into two groups—the red oaks and the white oaks. The acorns of these two groups differ in significant ways. The red oaks have more fat, a desirable quality, but also more tannin. Tannin is an astringent compound that reduces palatability and digestibility of plants. It also deters insect infestation and may serve as a preservative. White oak acorns have less tannin, but also less fat. They begin to germinate soon after they fall, while red oak acorns remain dormant until spring. The nutritional value of an acorn erodes quickly when it germinates. Preliminary research demonstrated that many animals could distinguish between the two types of acorns, but there had been no studies to determine if these acorn types were treated differently when squirrels were making storage decisions. Steele and Kropowski hypothesized that squirrels would give preference to red oak acorns for storage.

In their first study, they simply offered park squirrels an acorn from either the red or white oak group, and noted whether it was eaten or stored. Their findings showed that nearly all white oak acorns were eaten immediately and that most red oak acorns were buried. Their study also reinforced the observation of other squirrel observers before them; when white oak acorns were cached, the squirrels would, with a quick nip to the bottom of the acorn, destroy the embryo and prevent the acorn from germinating.

Germination is not the only threat to a squirrel's winter food supply. An acorn containing insect larvae may have little if any nutritional merit by the time a squirrel retrieves it. Our intrepid researchers gathered and x-rayed hundreds of acorns, and then offered them to squirrels. The acorns containing insect larvae were not cached but eaten, regardless of oak type. Larvae and all.

What cues were the squirrels using to determine which acorns were the least perishable? Steele, Kropowski, and now Smallwood, decided to make some trick acorns to further test the discernment of the fluffy-tailed rodents. The nut meats of white oak acorns were ground and varying amounts of fat and tannin were added to simulate the composition of different types of acorns. These nut balls were then reinserted into acorn shells. Some red oak blends were put into white oak shells, while white oak blends were put into red oak shells. Chemical compounds that might be used to differentiate red from white oaks were leached from the shells before reassembly took place. The squirrels ate them all. When the nut blends were inserted into shells that had not been leached, however, the squirrels cached the red oak shells regardless of the contents, and ate the meats in the white acorn shells. The cue must come from a chemical in the shell! In their next trial, they presented squirrels with typical dormant red oak acorns, and also with red oak acorns that had just started to germinate. Although there was no visible sign of germination, the acorns from the second group received the white oak treatment from the squirrels. Despite the technological arsenal of the human researchers—x-rays, chemical assays, solvents—they have yet to detect squirrels' cue, but suspect it is likely to be a plant hormone released during germination.

Their research suggests that red oaks and white oaks have separated in their dance with the squirrel. The red oak has developed a symbiosis with the squirrel that the white oak has not. Our team of squirrel biologists hypothesized that in a forest containing both types of oaks, the red would be more widely dispersed and the white oaks would be clustered. They counted oaks on forest transects and found this was the indeed case.

As you might guess, the red and white oaks have developed different parameters for germination. Red oaks do well in sunny, dry conditions. White oaks are more tolerant of shade and moisture. Has the white oak found that the risks of dispersal outweigh the benefits? In our region, red oak is far more abundant than white oak. Perhaps the red oak can attribute its success in part to Priscilla's clan.

When Priscilla finished her breakfast, she took one last nut between her teeth and headed back into the gale. With another bold launch into the snow, she bounded to a tall yellow birch and hoisted herself to the summit. With the branches thrashing in the wind I could see no way she could maneuver to the forest beyond, but I'm not a squirrel. When the wind velocity dipped, Priscilla scampered to the end of a branch and leapt. Paws spread, she hit the distant tangle of blowing twigs, managed to grab one, righted herself, and was off through the treetops to enjoy her hazelnut in the comfort of her nest. She tackled wintery weather with the innate competencies of her kind. I should have known she would. After all, squirrels have been shaped by snow and blowing branches as well as by oak trees.

Winter Weasels

~

COLD TEMPERATURES ARE BEST endured by those of some rotundity, those with a low surface to mass ratio. If you were to design a creature to endure the thermal challenges of winter, roly-poly would be a good basic shape to start with. Cold is not the only challenge posed by winter, however; finding sufficient calories is equally daunting. One of the most abundant, high energy, easily assimilated foods, available year round, comes in furry little packages that can be found living beneath the snow—those tiny mammals of celebrated fecundity—the mice, voles, and shrews. A creature able to enter that world beneath the snow would eat well indeed, and would have access to some of the warmest shelter available, too. Perhaps portliness is over-rated as a design for winter survival. Let us return to the drawing table and instead draft plans for a miniature carnivore, low and slim, that can enter the dark, close realm beneath the snow. Let us give this little hunter eyes that see well in the light and the dark, a keen sense of smell and good hearing, hearing that includes the high frequencies of small mammal vocalizations. Let us give it boundless energy and curiosity, and just for the heck of it, enduring playfulness.

You see where I'm going with this—we are, of course, re-inventing the weasel. Access to the caloric jackpot of tiny mammals stokes the weasels' furnace through the cold season, and provides the energy needed to travel to find enough to eat. Weasels eat about 40% of their body weight in food each day.

The winter of 2012/13 was an especially good year to be weasel-shaped, since it was also a very good year to be a rodent. After two years of abundant seed production and a fairly easy winter, the mice, voles and squirrels were able to raise many offspring. These fuelled a successful child-rearing season for the weasels. I have seen film footage of baby weasels. The eight fuzzy wrigglers in their nest of fur and feathers kept up a din of chirps, trills, and squeaks. Later they emerged from their den, leaping and wrestling and continuing the chatter. In the final film sequence, the young weasels set out to explore the surrounding territory, bounding off in tight formation like a tiny roiling river, pausing in synchrony every few seconds, all heads periscoping up on long necks.

Two weasel species can be found in Vermont. The long-tailed weasel, *Mustela frenata*, is the largest, with males about 16 inches in length and weighing about the same as a red squirrel. The females are a few inches shorter and typically weigh less than half as much as the males. The short-tailed weasel, *Mustela erminea*, is a dainty beast. A female might be just seven inches long and will weigh less than a chipmunk. Males are typically closer to eleven inches in length. Their paired footprints are so tiny, just an inch or two wide, that they can be mistaken for those of mice or shrews in soft snow. As the names suggest, tail length helps distinguish the two species; the short-tailed weasel's tail is about a third of the length of the rest of the body, while the tail of *M. frenata* is half the length of the rest of the body. In the summer both weasels are a dull, leaf litter brown with a white or cream-colored belly and a black tail tip. In the winter, both weasels don elegant white coats and keep the black tail tip. This white coat with the black tail tip is the fur known as ermine (ermine is also another common name for the short-tailed weasel). Ermine is the fur that adorns the raiment of royalty, snowy white accented by the black tail tips.

Of course if I were in charge of designing animals, I wouldn't make any carnivores at all. It is just as well that I'm not, for I admit it would be a less wondrous planet without such beasts as owls and bobcats and foxes. My ambivalence about carnivores has grown since I have developed relationships with many herbivores, among them Priscilla, the gray squirrel that I raised as an orphan a few years ago. She is a very regular visitor to my house and I take a keen interest in her activities and welfare.

One afternoon I saw the very fresh tracks of a weasel in my yard, and the scene of a kill, a large bloodstained depression in the snow. The tracks leaving the scene recorded the very short bounds and drag marks of a heavily laden weasel. Priscilla did not appear again for an unprecedented six days, but appear again she did. The weasel, it seems, had taken care of a moral conundrum for me by dining upon the very large rat that had moved into my hay shed. On a recent day of mild weather, as Priscilla and the other squirrels were celebrating the break from the cold, I saw a flash of white leaping across the yard. The weasel's tracks led over a steep bank and into a tunnel in the snow. While Priscilla lolled in a tree overhead, idly nibbling on bark and buds, I admired the weasel's neat footprints. When next I looked up, weasel and I were staring at each other from a distance of five feet. The weasel's black eyes and nose formed a neat equilateral triangle against the pure white of her fur, so alert, so poised, so beautiful. Here was perfection. The weasel did not find my appearance as captivating, little wonder. She wheeled and bounded back into a tunnel. As I stood guard beneath my squirrel friend, I thought about kings and queens with their vanities and human foibles, and knew that I had just seen the only creature truly worthy of wearing an ermine coat.

April Showers

~

THIS MORNING THE SQUIRRELS and chickadees arrived with their winter coats set at maximum
loft. They have just made it through one of the coldest
nights this winter has mustered. As deeply wintery as
this morning seems, I know that within a few hours the
direct sunlight of mid-March will make them sleek with
warmth. It is this knowledge that reminds me of another
image at odds with the polar cold; beneath this two feet
of snow and an unknown depth of frozen soil, a host
of amphibians snooze in yellow polka dotted pajamas.
They await their big moment of the year. If the March
sun has anything to say about it, the salamanders have
only a few weeks to wait. These black eight-inch long
spotted salamanders are of the genus *Ambystoma*, the
mole salamanders, denizens of the earth. Their ancestors
returned to this region with the temperate forests as the
last glacier trickled away far to the north. For the thou-
sands of years that have ensued, spotted salamanders
have prospered.

Although solitary for most of the year, spotted salamanders are not immune to the allure of spring, and when the first thawing rain reaches them, their fancies turn to thoughts of love. On those dreary rainy nights after the ground thaws and when temperatures are above 40° F, these salamanders begin the annual trek to their breeding pools. No obstacle is deemed insurmountable (although some might prove to be) as the lure of the pool and the other spotted salamanders they will find there exerts its pull.

Once in the water, buoyant and convivial, the salamanders assemble for the Ambystoma Ball. The males generally arrive before the females, but they aren't shy about getting the dance started. I understand that on a year when the salamanders synchronize their arrival at the pool perfectly, their courtship dance is spectacular. David M. Carroll, in his book *Swampwalker's Journal*, describes the scene:

> "It seems that all of the salamanders I have been looking for all spring are here, and have all become one, in a mesmerizing black mass of interweaving sun yellow spots... a great communal congress of salamanders continually weaving among themselves in a dense, nearly spherical mass.... Limbs tucked against their sides, one main stream of salamanders slips from one pole of the rough globe to the other, while others slide around in all directions.... Adding to the magic, the *perpetuum mobile* of the entire living orb remains stationary in the water."

I have never witnessed such an extravagant display, although I have seen more subdued variations. It could be that Vermont salamanders are simply less flamboyant than those in New Hampshire, but that seems unlikely; the evenings when these balls occur, I am engaged elsewhere, as I shall soon reveal.

These courtship rituals typically occur in vernal pools—temporary wetlands that hold spring rain and snow melt. In this habitat, amphibian eggs are safe from predation by fish, but each year the young develop in a race with the sun as the pools dry. The salamanders' longevity (twenty years or more) assures the survival of their genes even if developing young sometimes lose the race.

This year the salamanders will once again head to the pools. Superimposed upon this ancient world, however, is a new world of houses, shopping centers, roads and cars. In areas where salamanders are forced to cross wide, busy roads, populations of these animals are likely to disappear. What about populations of amphibians that must cross even moderately traveled rural roads? For creatures with legs less than an inch long and blood as cool as the April night, a simple road crossing is a dangerous prospect. Fortunately for some amphibians, on such nights another colorful spectacle blooms along roadways in our area—groups of people in raincoats and reflective vests wielding flashlights. These are kind-hearted folks who turn out in *amphibian weather* to see some of these handsome salamanders and to help them safely reach their pools.

One of my most rewarding and stressful tasks is to keep track of where and when these amphibians are likely to be moving so I can make sure the amphibian escorts are in position when they're needed.

Those of you who roll your eyes and think we must be crazy are partly right. Chances are good that most of us are a bit crazy. However, these crossing brigades may prove critical to the survival of amphibian populations as traffic on our roads increases. The likelihood of an amphibian making it across a particular road can be calculated by knowing the speed of cars, the number of cars likely to pass, and the amount of time an amphibian spends crossing the *kill zones*, those parts of the road that tires travel. A study modeled the number of spotted salamanders in western Massachusetts that would likely need to cross roads to reach breeding habitat and the percentage of the population that would be killed by cars in the attempt. This study projected a gloomy future for the spotted salamanders of western Massachusetts. In their conclusion, study authors Gibbs and Shriver state that, "If efforts are successful in limiting rates of traffic-caused morality to less than 10% of individuals attempting to cross roads during their migration circuit to a particular pond, e.g., by tunnel construction, road closure, or physically transporting individuals, then those efforts are likely warranted to stave off local population extirpation."

~

It is the spotted salamanders that steal the show on rainy spring nights, but they are not the only amphibians that benefit from our help. Wood frogs also migrate to breeding habitat on those rainy spring nights.

One night, with temperatures hovering close to 40° F and a light rain settling on top of dry soil, not conditions to excite a spotted salamander that early in the season, I headed to a crossing site that I knew would be hopping—wood frogs are less discriminating about temperature and moisture than salamanders. In fact one of their charms is how very indiscriminate they can be.

The more enthusiastic of the male frogs were easy to spot. They perched on the asphalt in an upright posture. With their dark bandit masks they looked like miniature highwaymen hoping to plunder a stagecoach. Others, spent from their travels, had no energy left for good posture. They felt like empty frog sacks as we scooped them up. I hoped the pool party would rejuvenate them. Some of these seemed grateful not so much for the lift across the road, as for a warm spot to rest. They hunkered into my hand and seemed reluctant to leave.

Soon the nearby pond quacked with the mating calls of the males. As we walked the road we would sometimes hear the cluck of a frog warming up for his performance as he hopped through the woods. A few offered a little *kkRRruuK* when I picked them up. Still others were so eager to mate I had to pluck them from my fingers to set them free once we crossed the road.

I thought I would leave at 10:30 when the rest of the team headed home. I don't know if it was the abrupt increase in the volume of rain or the reduced activity of the human patrols, but in a moment I found myself in the midst of a flood of frogs. They flowed down the banks and into the road, a torrent of amphibious life. I couldn't begin to keep the road clear, so I was very grateful that the couple of cars that passed were driven by sympathetic folk who were happy to have a path cleared through the parade.

We saw just three spotted salamanders that night, but I was not disappointed. While there, we managed to nearly eliminate mortality. I knew we had helped over 400 frogs reach their breeding pool in safety.

Fever

~

THE FIRST WARM DAYS OF SPRING are among the sweetest of the year. One does not stay indoors willingly. Imagine how you'd feel about such a day if you had spent the winter in a dark hut of mud and cattails with a bunch of damp muskrats. If I were a muskrat, I'd certainly be ready to stretch my legs, have a little fun, and put as much distance between myself and my housemates as possible. Judging from the muskrat remains that sometimes appear on the roads in spring, they concur. Such might have been the fate of a muskrat I met while out on the town one fine spring night.

I had just seen the early show at the Latchis Theater and was starting to walk up Main Street when I noticed a group of nervous, excited young people staring into the darkened entryway of a jewelry store. I investigated. In the corner huddled a muskrat that seemed to have had an overdose of excitement. As tires sped by in ominous proximity, I knew this muskrat's chances of a safe return to the river were only fair. Fortunately I was wearing my brown tweed coat, a versatile garment that can serve as a small animal net and tranquilizer. The young people stepped back as I removed my coat and explained my intentions. As the coat descended, however, the muskrat revived and took evasive action. Hugging the edge of the building, she scuttled down the street and turned in at the theater where a long queue had formed for the next show. Those in line were so distracted by the advance of the large tweed coat that they didn't notice the muskrat until its paws scrabbled over their feet. I was able to track her progress by the shrieks from the crowd. She finally took refuge under a bench in the lobby. When the crowd dispersed and only a small group of curious people remained, I persuaded the muskrat to leave her refuge. I cornered her by the popcorn machine. The wool coat had the desired effect and she relaxed for her transfer to a cardboard box. Among those assembled was someone who lived near a wetland complex that sounded like muskrat heaven, and she offered to transport the muskrat to this destination. I hope the muskrat found the habitat suitable and wasn't too disappointed to have missed the late show.

My conversations with the people who gathered made it clear that most of them had never encountered a muskrat before. These good people may be excused for wondering if muskrats eat chickens or if they are, in fact, a type of rat; except when enjoying the delirium of spring fever, muskrats frequent habitats infiltrated by few humans.

Although less visible, muskrats have a great deal in common with beavers. Like beavers, muskrats have dense, lustrous, brown fur, large hind feet, and lips that can close behind their incisors so they can chop vegetation under water. They also produce a pleasant-smelling musk that is used to convey information to other muskrats. Like beavers, many muskrats build dome-shaped houses and excavate canals. Muskrats and beavers also excavate tunnels and dens in the banks of rivers.

The two species can be readily distinguished. Muskrats weigh between two and five pounds, an adult beaver weighs about forty pounds. The muskrats' tail is more like that of a rat than a beaver, but is slightly flattened laterally. The tail is visible even when the muskrats swim, cutting serpentine ripples behind them.

The cattail is the plant most venerated by the muskrat, for it is palatable in its entirety, from its abundant starchy rhizomes to its flowers. The cattail can afford to spend so little energy on chemical defense against herbivores since it proliferates in an environment where very few plant-eaters dare to tread. There is no place a muskrat is more at home. Cattails grow in saturated mucky soils, and are well adapted to fluctuating water levels and standing water. In the right conditions it takes only a seed or two or a raft of dislodged rhizomes from upstream to establish a dense colony of cattail clones. When a muskrat takes up residence in a cattail marsh, it begins to make improvements. Although muskrats are perfectly capable of clambering through the vertical jungle of cattails, they create a maze of trails and canals for unimpeded travel.

A muskrat builds a home by constructing a platform of mud, cattail rhizomes, and whatever else it excavates from the surrounding marsh bottom. Once this platform is above the water level, the muskrats pile plant material into a dome on its surface. They then excavate entrances from below the water and nest chambers in the pile above. Muskrat and Beaver are like the first two of the Three Little Pigs. While Beaver makes a house of sticks (moderately impenetrable), Muskrat makes a house of straw. It might take more than a huff and a puff to blow it down, but it would be a simple matter for a hungry coyote to dig its way in. Fortunately, coyotes are deterred by the standing water, dense cattails, and deep muck that must be endured to reach a well-situated muskrat village. When winter freeze-up creates easy access to the marsh, the mud mixed with the plant materials in the muskrat lodge will have frozen as hard as stone.

In the process of creating platforms for building and feeding, muskrats create deeper open pools within the cattails. When freeze up comes, the water must be deep enough that the muskrats will be able to swim and feed beneath the ice. While this is clearly important if you happen to be one of the resident muskrats, it is also important if you happen to be a black duck or a water hyacinth. In the case of waterfowl, the pools provide a sheltered haven. In the case of emergent marsh plants, the muskrats keep the cattails under control and increase the herbaceous diversity of the marsh.

If the movie theater muskrat was female, she will likely spend her summer raising a couple of litters of kits. Such muskrat families seldom stray far from their lodge. Mother muskrats will defend the surrounding territory vigorously. During a year with a high muskrat population, female muskrats will be quite visible and audible as they protect their family feeding ground.

All territorial animosity is forgotten when warmth becomes the priority. Ten to fifteen muskrats have been found sharing a lodge in the winter. Compared with the exposure many of our resident birds and mammals endure, the companionable winter quarters of a muskrat must be downright cozy. Still, the arrival of spring must be welcome indeed.

The Importance
of Being Otter

~

LAST SUMMER I was growled at by a bush. Over the years, I have been growled at by cats, dogs, chickens, and once by a hay bale, but never before have I been aggressed by a shrubbery. I admit that the warning was not unprovoked. I had, after all, very deliberately pushed the bow of my canoe right into said shrub in an effort to photograph the shrub behind it. The growl, a deep ominous rumble from the vicinity of the roots, gave me pause. How should I proceed? Was the bush really growling at me? The only way to be sure, I thought, was to advance a little further. Yes, in fact, the growl did recur, and with a more menacing note. This time I retreated rather quickly, and then let my mind wander through the possibilities. Probably not the bush. An alligator, then? I considered an alligator unlikely at Branch Pond in the Green Mountain National Forest. A bear might growl like that, but a bear would have retreated inland. I decided on an otter. The sort of otter most likely to hunker down in a shore-side den and growl would be the sort that had a litter of kits to protect.

Not wishing to disturb the growler further, but hoping to confirm my guess, I approached the shrub from the shore to look for evidence. Sure enough, I could make out a dark chamber in the bank beneath the bushes, and about thirty feet away I found an impressive otter latrine, sparkling with fish scales.

I wished I could stay and hide on the opposite shore to watch the otter show that evening, but duty summoned. I could only imagine the scene of otter mother escorting her family out to fish the trout-stocked waters of the pond later in the day.

As much as I enjoy my evenings with the beavers, it is my secret hope that I will have a chance to watch river otters. Beavers are stolid creatures. Otters, by all accounts, exude *joie de vivre*. Most of what I knew of them I have learned following tracks in snow and from nature documentaries. This evidence, however scant, marks river otters as voluptuaries—connoisseurs of sensation, fans of frolic.

This quality, disparaged in such instructional tales as *The Ant and the Cricket*, gives otters high honors on my merit scale. Here's why: There are days, especially bitter cold ones in winter, when I feel the weight of the suffering inherent in the lives around me. I acknowledge that the life that has arisen on this little planet wouldn't be here at all without such unpleasantness as predation, and the fear that accompanies a will to stay alive. One day, while brooding on this unfortunate fact, I decided to cultivate joy and appreciation as a counterbalance. As something of a rational empiricist, I admit the notion is overly metaphysical. Still, I can't help but feel it would be a tragedy for such a wondrous conglomeration of life and landscape to have arisen here and to go unappreciated and uncelebrated. By this principle, any creature that savors a good roll in the snow is making the world a better place.

I have followed many otter trails in snow. These animals don't miss an opportunity for a belly slide. While usually found along a river or stream, some of the best otter slides I've seen have been down mountainsides, as the otter moved from one watershed to another. I have seen places on steep riverbanks where the otters slid down the same bank repeatedly.

This recreation has been noted by many keen observers of the natural world. John James Audubon recorded a group of otters sliding down a slick mud bank "twenty-two times each" before they were spooked by his audience. Ernest Thompson Seton describes this sliding behavior as "... the only case I know of among American quadrupeds where the entire race, young and old, unite to keep up an institution that is not connected in any way with the instincts of feeding, fighting, or multiplying, but is simply maintained as an amusement."

Alcott Smith, one of the foremost observers of wildlife and their behavior in the Northeast, told me about a baffling sound he heard on a late season canoe trip. A skim of ice covered the river, and from somewhere came the sound of splashing and shattering. He and his companion thought it must be a moose wading. Instead they found a family of otters, a mother and her offspring, swimming up beneath the ice and breaking it with their heads. Alcott explained that otters' skull and musculature are adapted to allow them to become battering rams to maintain openings in ice through much of the winter. The activity he witnessed may have been good practice, but it seemed to him that these otters were having fun; and the fun lasted for more than an hour.

When the ice sealed my beaver pond, and my evening vigils took a winter hiatus, I was sorry that my hope of watching otters frolic in Popple's Pond would have to wait. In mid November, I went to a conference at the new museum of Adirondack natural history, The W!ld Center, an experience that helped tide me over. The museum features huge aquariums that display the life of different types of water bodies found in the Adirondacks. Naturally, it includes a rushing rocky river exhibit. When the museum is open, I was told, this display includes otters.

In late afternoon, when I was playing hooky and enjoying the museum, the lighting changed and water began gushing from openings high on the walls. The rocky river tank now had a waterfall. I sat down on a bench and hoped. Sure enough, a minute later a river otter slipped into the water and sped toward my seat. She ricocheted off the side of the tank in front of me, dove under a log and, trailing bubbles, flew to the far side of the tank. Here she did a back flip and returned. Soon there were two otters and I watched, mesmerized, for an hour. Their default position in the water was floating like a log, though this seldom happened, for they were in constant motion. They didn't swim with a single stroke style; for high speed in open water they undulated like porpoises, even leaping above the water, and frequently pushed off of objects in the water to boost their speed. When maneuvering through obstacles, or wrestling with each other, the sinuous bodies and appendages operated independently to propel them. They didn't take any notice of me, it seemed, although they often lingered in front of my bench. Then one of the keepers arrived. "Hi Squeak! Hi Squirt!" she announced. The otters' heads popped up. The keeper then dashed around the tank, turned, and ran back the other way. The otters pursued her ecstatically, ever alert for a change in direction. When the game finished, I asked the young woman about the lives of these otters.

Both females, these otters had been orphaned and

raised by people. Not only were these otters not afraid of people, they had some suspicion that they might be people, a delusion likely to cause misunderstandings if they were released. I arranged for a tour of their backstage accommodations. They had a large enclosure with a tank, dirt floor, many toys, and climbing and sliding equipment. The otters talked eagerly to us with little growly grunts. Yes, I could imagine a bush sounding like that.

Later that day, when the otters were in the exhibit again, I sprinted around the tank. Sure enough they chased me, too, always a bit ahead of me yet still in pursuit. No planet with otters can be entirely gloomy.

Charmed, I'm Sure

⁓

A FRIEND ONCE ASKED, "Which do you find more appealing, birds or mammals?" Since I find nearly all animals, from slugs to gazelles, appealing, any sort of ranking is not to be undertaken lightly. My favorite animals are alert, curious, interactive, and expressive. I admit that for much of my life my partisanship has been with the mammals, but on the day when the question was put to me, I voted for birds. I tip my hat to Benjamin Robin for opening my mind to the charms of the avian class.

That summer I had helped Fred Homer, the local wildlife rehabilitator, with some of his charges. His tutelage would provide some of the training I needed to get my own wildlife rehabilitator's license. When Fred asked if I'd be interested in taking on a robin, he warned me that this was one of the sorriest looking birds he'd encountered. The day I met Benjamin Robin, I had to agree with Fred's assesment; this young robin was indeed a peculiar sight. The missing feathers on the front of his head gave the poor bird a masculine appearance, and I found myself calling him Benjamin. His remaining dull, marred feathers indicated that he had suffered from illness, stress, or malnutrition. One wing drooped and one leg gimped, but two alert dark eyes scanned the world with interest.

The person who brought the bird to Fred said that the robin had hopped over to him as if begging for a handout. Fred's best guess was that this bird had been hand-raised, but had suffered one or more of the mishaps that often occur when inexperienced people try to foster baby birds. Benjamin didn't seem a likely candidate for a long life in the wild—it was possible that he would never fly, yet he seemed so vital that I had to give him a chance. With little fear of humans and with such damaged feathers, this bird would need an unusual strategy for rehabilitation. Most birds will begin to assume independence shortly after being released. They will continue to beg for food at first, but return less often to be fed each day, until, usually within the first couple of weeks of freedom, they no longer return. With Benji, I decided to take advantage of his bond with humans to keep him under close supervision until his new plumage grew in sometime in the early fall. In the meantime I would do my best to teach him the skills he would need to be a robin.

Flight was to be the first challenge. My initial attempt to encourage flight failed. When I released Benji from a height of a couple of feet he dropped with no fluttering. Still, one could see that he expected more. He often paused in his pedestrian explorations and looked skyward with a puzzled expression, as if he knew there was some way to get up there, he just couldn't remember how. Since I had no way of knowing how to go about flight myself, I could only offer encouragement. Whenever I caught Benji pondering the heavens, I would earnestly assure him, "You can fly. Go ahead. You can fly!"

This strategy proved to be the right one. His first efforts at flight powered the hop that freed him from his temporary aviary in the mudroom. Soon he was able to move from exploring the mysteries of the cracks between the kitchen floorboards to subduing the creatures that dwelled within the rush seats of the antique dining chairs. My unreasonable housemate banished the bird from the dining area, so my office became Benji's training center. After many short sorties and crash landings over the course of a few weeks, he managed to fly twenty feet from my desk to a chest in the hall. I knew then that flight, at least, would return.

Another skill this robin needed was fluency in his native tongue. Robins have a great deal to say to each other, including the melodious singsong performed by males to stake out their territory and woo a mate. Young birds must learn this song, and the most skilled vocalists are the most likely to win the affections of a female. Although Benjamin's sex was in question, learning the robin song seemed important, so I often played him recordings of robin vocalizations. He would perk up and react to the different calls, preening his feathers contentedly when he heard conversational clucks, chortling along softly to the robin's song, and hunkering down when he heard the high-pitched warning note.

I was surprised to discover that he enjoyed other music, too. He even liked my singing. I made a playlist of Benji's Top Hits that I would leave playing for him when I left him alone. From other parts of the house I would often hear him joining in the recordings at the most compelling parts—always in harmony, or so it seemed to me.

It wasn't just singing Benji appreciated. He assumed an expression of interest when I talked to him. He always had comments of his own to share. He soon trained me to speak in *robin English*, a ridiculous lilting speech that produced the highest level of attentiveness. What must the neighbors think?

While birds are more limited than many mammals in the range of facial expressions they can make, they can be quite expressive in other ways. Benji greeted me by flitting the tips of his wings and hopping about with animation. He often settled down close to me, or on my lap, shoulder or head. With breast feathers fluffed he would chirp softly with each breath like a contentedly purring cat. He did not like to be touched, however, and would jab his gaping beak defensively at hands that approached too fast. He raised the feathers on the top of his head when he was alarmed or annoyed.

The computer keyboard was a source of conflict. My hands spent so much time pecking there that Benji naturally assumed there must be something interesting inside. While bird poop caused no harm in other parts of the room, if cleaned up quickly, moisture on the keyboard could be catastrophic. Benji learned that a sprint across the keyboard brought excitement to dull moments.

Baths were a jubilant occasion, and often indulged in. Following a splash he would sprawl to dry in Sybaritic bliss beneath a lamp, one wing spread and tail fanned, his neck stretched out, shifting occasionally, raising and lowering feathers to expose every surface to the delicious heat.

Worms, on the other hand, were alarming. I provided him with a pan filled with soil and earthworms purchased from the bait cooler at the gas station. When Benji first felt compelled to peck at a burrowing worm, he leapt back in horror when the writhing pink serpent was revealed. The worms lived in peace for couple of days as the robin fluttered wide around the pan, glancing at it with grave suspicion. Gradually, however, the allure of the sound would draw him back, and each time he grew more bold. Within the week he became the Dread Robin of earthworm lore, and it was difficult to keep him in worms.

Once Benjamin could fly and had conquered his fear of food, I began to let him out. I hoped that he would acquire the skills needed to forage as a free robin, but that the bond we forged would keep him nearby  where I could offer some protection until his new plumage came in. And so it came to pass that Benjamin Robin discovered the treasures concealed beneath leaves on the forest floor and the view from the roof of the house. He came and went frequently from my office window, and roosted on top of my desk most nights. When I was outdoors he would chatter away to me, and would often fly over to ride on my hat.

It was in September that Benji's transformation occurred, with scarred drab feathers falling out one by one, to be replaced by the plumage of a healthy adult robin, and a female at that! It was also September when the waves of robins started sweeping south through the woods. How would Benji respond? One day in early October she was seen sharing her dish of worms with another robin. Two days later, she was gone. I like to think she was on her way, drawn south by the companionship of a passing flock and by the mysterious compulsion that drives such journeys.

~

I suppose I should blame my college ornithology text for my surprise that a robin could be so, well, personable, more so than many mammals of my acquaintance. This book instructed that structural differences between the brains of birds and mammals resulted in different behavioral capabilities. Mammals have a "high capacity for learning," while "the behavior of birds is largely mechanical, stereotyped, and instinctive." Our understanding of avian neurobiology, however, is in the throes of a revolution. Studies have revealed that parts of the avian brain that were believed to be primitive, process information in much the same way as mammalian brains. This comes as no surprise to the researchers who have been documenting learning in birds that even our closest primate relatives can't duplicate.

Among these researchers is Irene Pepperberg. Her now famous studies with the African grey parrot Alex rewarded her confidence in bird brains. Alex could tell her how many (from none to six) of an assortment of objects were a particular color, shape, or substance. He could recognize and ask for over forty different objects.

He could negotiate with his trainers when he wanted a different object than the one being discussed.

The crows of the Pacific island of New Caledonia make tools from twigs and leaves that they use for extracting food from hard-to reach places. They carry these tools with them from place to place, and pass along design improvements to their associates.

In a study with a group of New Caledonia crows, researchers deviously placed a favored crow food at the far end of a pipe. This end of the pipe was capped, but had a small hole in it. The crows had an assortment of dowels, none as long as the pipe, with which to attain the prize. Unstymied, they demonstrated their spatial problem-solving skill by poking a dowel through the hole at the capped end, pushing the food up the pipe, and then moving to the other end to pull the food the rest of the way out. In another study, a captive New Caledonia crow named Betty faced the puzzle of how to lift a bucket of food from a well. Behavioral ecologists provided her with two pieces of wire—one hooked and one straight. Before she had a chance to act, a fellow crow flew off with the hooked wire. Betty (who had never met a wire before) simply fashioned a hook from the remaining piece and lifted the bucket. This might not be rocket science, but these crows are the only non-human animals that have demonstrated the ability to fashion hooks from unfamiliar materials to solve problems, and to improve tools and share their improvements. Not bad for bird brains.

I was able to trot out these studies when I argued my case for the appeal of birds, but how a creature scores on an I.Q. test is only a part of what makes it attractive to me. I dare say I would find birds less appealing if they devoted their energy to designing bridges or debating the merits of different diets. Their achievements in aerodynamic flight, migration, and survival on this competitive planet inspire awe.

My relationship with Benji transformed the way I see all birds. When the trees filled with returning migrants the next spring, I found myself absorbed by their activities. A few returning robins seemed quite interested in Benji's Top Hits, but if she was among them, she had outgrown her interest in human company.

Eloise, on the other hand, takes a great interest in everything I say, especially if I emphasize the melody of the speech. This orphaned robin arrived a week ago. Her eyes were still closed, tufts of down sprouted from her pink head, and fans of tiny pin feathers sprouted from her wings. By the end of this week I expect she'll be flying. By the time she's seven weeks old, she'll be ready for independence. Already she has added a bit of beauty, grace, and charm to the universe.

The Awesome
in Possum

~

HERE IS WHAT MANY PEOPLE KNOW of the Virginia opossum: the pale shape scurrying awkwardly through the dark towing a pale rat-tail; the hissing alligator maw of a startled opossum; the numerous lifeless lumps of fur along roadways. These people wonder at my inordinate fondness for possums. Here is what I know: When I carry Alexandria and Samuel Gompers into the office in their carrier and plop them on the couch, they wake up and yawn and sniff the air hopefully. I feed them some grapes, which they chew thoroughly with their Muppety mouths, eventually spitting out the skins. They will then ramble about the office a bit to see if any interesting smells have materialized overnight before crawling back into their den to sleep for the rest of the day. I peek in often since they sprawl in amusing postures, and besides, they like to have their bellies rubbed. At almost a year old, Alexandria is beautiful, the sooty fur around her eyes set off by her white plush coat.

Her brother is a great bull possum. His head and neck are broader than his sister's and the sides of his mouth tend to droop from all of the salivating and lip-smacking that are an important part of male possum scent-marking and courtship. Both are good-natured, docile creatures who, like all possums, seem designed for life on a different sort of planet, one on which life happens slowly, where temperatures are warm and predators few. Still, somehow opossums not only survive on our planet but they have been slowly expanding their range.

They succeed by reproducing prodigiously. A female opossum can raise up to thirteen joeys at a time (though they seldom produce that many), and typically have two litters a year. They also succeed by eating almost anything, the smellier the better.

Samuel Gompers, Alexandria, and The President were tiny and pink and smelled of carrion when they were delivered to my custody after a caring motorist removed them from the corpse of their road-killed mother. Their weights straddled the 20-gram mark, below which chances of successfully raising possums are considered near nil, indeed a fourth sibling did not make it through the first week. The other three, however, transformed from decidedly homely into cute little fluffy joeys. Like all proper baby possums, they joined me, their surrogate mother, on some of my daily rounds, tucked into a modified satchel. By then they were already prodigious grippers and would ride along clinging to the inside of the *porta-possum* with their heads poking over the top to survey the passing world.

The possums came with me to beaver camp and began their first independent sorties in a canopy tent there. Opossum families make a sneezing noise to keep track of each other, and I did my best to imitate this sound. The three wanderers always kept up a chorus of sneezes and stayed within range of mine.

Because speed and agility are not hallmarks of the species, I did not notice until too late that Samuel Gompers did not walk normally. Opossums are particularly susceptible to a type of bone disorder. Alexandria also had the condition, but to a lesser degree. I decided to keep them over the winter to see if the condition would improve. Now, as they approach their first birthday, it seems their destiny will be as ambassadors—to grow old sharing the wonders of possums with Bonnyvale Environmental Education Center campers and school children.

And wonders there are! The Virginia opossum has many distinctions. It is, of course, the only marsupial in the United States. A mother opossum gives birth to embryonic infants after a thirteen-day gestation period. About the size of a kidney bean, but with well-developed front legs, the newborns drag themselves up the short path to their mother's pouch. There they play a life-or-death game of musical nipples. A female opossum is equipped with thirteen nipples, and may give birth to twenty joeys. The newborns must latch onto a nipple in order to survive. Those that fail are out of the game. The joeys will remain attached to their nipple for about sixty days and are fully weaned a month later.

Opossums also have prehensile tails, play dead when threatened, seldom contract rabies, and are immune to rattlesnake venom. Here is the reason you will want as many opossums in your neighborhood as possible— opossums may be your best allies in reducing the incidence of Lyme disease. Research published in *Nature* (Dec. 2010), compares white-footed mice and opossums as Lyme disease vectors. The researchers found that mice groomed off and ate 1,021 deer ticks per hectare, but almost as many ticks, 906, fed on the mice and became infected with the Lyme spirochetes. Virginia opossums, well equipped to pluck off ticks with their iron-grip paws, attracted and ate 5,487 ticks per hectare. Of the 199 ticks that bit a possum and got away, only five carried the Lyme spirochete.

I now have eighteen little joeys—that's right, eighteen, and you should see the pile of them, snoozing away in their nest, fists clenched, twitching as they dream their possum dreams. They represent two litters, one of thirteen and one of five, all survivors of automobile encounters. They show great potential as decorative-but-shy tick magnets and sanitation workers. They are about eleven weeks old now, and should be ready to release by the end of summer. Want a possum or two for your backyard?

Old
One Eye

~

MOST WILDLIFE REHABILITATION involves raising orphans, a job made easier by the eagerness with which hungry, frightened young creatures accept any surrogate. Injured adults are different. Captivity induces so much stress that many wild adults injure themselves further trying to escape. When I got the call about an opossum that had been hit by a car, I feared she would be such a patient.

The possum was a sorry sight—dirty, disheveled, and with dried blood on her nose. Once she had a chance to become accustomed to her new surroundings, I gave her a physical exam. I found bruising and scrapes on one side, and a badly damaged eye, but no broken bones. She tolerated the prodding with remarkable equanimity; she did not try to wriggle away, growl, hiss or bite. She lapped some nourishment, pain medication, and antibiotic from a large syringe, and allowed me to wash her eye.

The tips of her ears and tail had been lost to frostbite,

proving that she had survived at least one Vermont winter. When she yawned, I could see that her teeth were rounded from wear. She was an old opossum! I would think it remarkable for an opossum to live long enough to become aged, given the myriad hazards they face, except that opossums often begin to age at a mere two years of age and seldom live beyond three and a half years even in coddled captivity. One expects tiny, active animals with a high metabolic rate to burn through life quickly; the hyperactive short-tailed shrew, for example, lives a maximum of 2.5 years, and has a metabolic rate ten times faster than opossums'. Why wouldn't opossums live as long as raccoons? These masked omnivores have a metabolic rate about twice as fast as opossums', yet they can live for twenty years in captivity.

Twice each day I treated Old One Eye, changed her bedding, gave her food and water, and each day her condition improved. In a week she was ready to go into an outdoor enclosure to develop strength. It was clear that her damaged eye would be lost, but opossums do most of their navigating with their noses and whiskers, so a one-eyed opossum has only a slight handicap. Two weeks after her mishap, the old girl was ready for liberty.

Lydia, the woman who rescued the possum from the roadside, suggested that her own home would be a good release site since it was near where the opossum was found, yet well away from busy traffic. When the day arrived, I drove up a long, winding, quiet dirt road and found Lydia's farmhouse surrounded by wonderful habitat. I carried Old One Eye's crate to the edge of a wetland thicket and opened the door. How her whiskers twitched as she inhaled the aromas of the earth! She strode to a small patch of muddy open ground, took a bite of the holy stuff, chewed a few times, and then spit it out again. She then wandered back up to where I was sitting and to my surprise climbed up into my lap. After gazing at her new surroundings, she slipped back down and wandered off into the thicket and into the life of a proper opossum.

Meanwhile, back at the ranch, the eighteen orphan opossums I had been raising continued my opossum education. I reported in the previous chapter that I communicated with young opossums by imitating the call that they make to each other, the group-cohesion-sneeze-sound—chh! chh! I recently read, however, that the sound mother opossums make to gather their brood is a different one—the lip-smacking noise I hear male possums produce when making overtures to females. When a possum mother calls, she is letting the kids know that she is ready to move along and if they plan to come they had better jump aboard. Once they are too big to fit in her pouch, they ride clinging to her fur. If they miss the summons or fail to hang on, they will be left behind.

I found that when I made the sneezy noise, the possums continued with their activities. If I made one or two of the soft lip smacking noises, though, I created an opossum stampede. The first time I tried this I was bringing the little pups their dinner. I was fresh from the shower, and my hair hung around me in long, dripping tendrils. As soon as I made the noise, an army of little opossums was climbing up my hair, hand over fist, on an earnest mission for the summit. Yes, it pulled like the dickens, but I couldn't stop laughing long enough to cuss at them. They climbed back on faster than I could detach them. I still can't explain how I managed to get out of the enclosure with no possums on board and with at least half of my hair still attached.

The possum joeys weigh about a half-pound each now. They sleep in fleece slings suspended from a horizontal pole. It is not unusual to find all eighteen of them piled into one small hammock, the great basketball-sized bulge hanging lower each day. Soon they will be ready to begin their lives as full-fledged wild possums.

I am grateful to all of my wildlife charges for allowing me to enter their nations in some small way. They answer questions and raise new ones. I now know how to summon a herd of baby possums, but I may always wonder: did Old One Eye climb into my lap to test the air from a different level, or was she seeking a moment of familiar contact before heading out into a new place—an opossum good-bye?

Roo, the Real Rabbit

~

As pets go, my rabbit is more decorative than interactive. When I return from a day of work, Roo does not greet me with leaps of joy. No, he continues contemplating the world from one of his two postures, sitting up (the *lump* position) or flopped on his side with his hind legs sticking out behind him. With floppy ears, immense dark eyes set into deep tawny fur, and a flattened face, he looks like a discarded plush toy. Only the little chevron nose that flutters above his nostrils betrays his living state. Roo does greet me on seemingly random occasions, hopping in circles around me while issuing his only sound, a low, pulsating hum. He makes few demands—no whining, barking, or beseeching stares. He sometimes nudges my ankles when it occurs to him that a massage would be welcome. When I rub behind his ears and under his chin, he settles down and closes his eyes in bunny bliss. When I massage his back and sides, he returns the favor and licks my hand. Indeed, grooming is one of his few activities, though I fail to see how it contributes to cleanliness. At such times, the odd little fellow sits up on his haunches, turns his stumpy forepaws up, licks them, and then rubs his nose, face and head. He then pulls each floppy ear over and licks as much of it as he can reach. Friends and I surmised that his calm outer life reflected a dull inner one.

This notion was not shared by Anne, my friend and bunnysitter. She believed that his large head contained more than fluff. At the end of July I left Roo in her care for a few weeks. The day after I dropped him off, Roo slipped through a small hole in her fenced yard. She hadn't seen him for several hours when she notified me. The bunny had disappeared into rabbit paradise. Anne's yard is a maze of pathways among islands of dense and varied vegetation. For hours I strolled the paths, peering into every place I thought a bunny might hide. I then enlisted Maggie, a dog of Roo's acquaintance, to aid in the search, to no avail. Neighbors were alerted, pictures of Roo were posted.

I wondered whether domestic rabbits have any homing ability. I found nothing authoritative on the internet, just one tale of a homing rabbit in England. This rabbit was sentenced to life in the wilderness for offenses that included book-eating and nibbling the mustache of a sleeping guest. Five times he was transported. Five times he returned. If this story were true, Roo could be attempting to hop the several wooded and hilly miles between Anne's house and my own.

Each day my optimism faded. Then the cats in Anne's neighborhood began to disappear. I fought disappointment with black humor, saying that Roo was to blame, but in truth I gave up hope for a bunny after the first two cats went missing. Coyotes and foxes were often heard and seen in this neighborhood. Although I consider loose pets fair game for any wild predator, I shuddered to contemplate Roo's probable fate.

Roo had been gone just a few days shy of a month, so when Anne called I assumed it would be to talk about other things. Instead, she reported that her neighbors had spotted Roo warming himself on their driveway at dawn! I drove right down. Her neighbor showed me where they had seen the rabbit. He had disappeared over a steep bank grown high with brush and poison ivy. That was the first of several days of rain, yet I clambered into the damp thicket to look for Roo.

I was standing on a rise partway down the bank when Roo detected my looming presence. You should have seen him! Up the bank he flew in mighty athletic leaps. I saw no more of him that day. The next day heavier rain curtailed the search. The neighbors' children watched their cardboard box-and-stick trap. I left a few jellybeans in his carrier near the place where he had last been seen. When the rain abated the next day, Anne and I resumed the search. From a perch on the bank I watched Anne wade through the brush below. She saw him first, and gestured. There was Roo, noble and alert, with the air of a competent wild creature. As I moved slowly toward him, my old companion bolted. I never dreamed he could move that way, and command so much territory. The terrain that had me clambering awkwardly hindered him not a bit.

I found him crouched in a patch of tall grass. I sat down a few feet away. Over the next twenty minutes, I watched him relax and transform from Roo of the Wilderness back to Roo of the Carpet. When the transformation was complete, he hopped over to me. I grabbed him, an act that once would have provoked panicked indignation, but this time he seemed relieved.

I wondered if Roo would be content to return to his life of domestic simplicity. A few days after his return, I opened the door to the wide world. If Roo wanted to be a wild bunny, I would let him. Roo hopped out and explored his old familiar lawn with animation, rubbing his chin on plantain stalks and following the squirrels. After several minutes, he hopped back into the house and flopped over onto his side with his legs sticking out, his little chevron nose fluttering above his nostrils.

Bean

~

T<small>HE ORPHANED RED SQUIRREL</small> had arrived a
week before, chilled, hungry, and dehydrated. Now
bright-eyed and fuzzy-tailed, Bean attempted to scram-
ble off the scale while I attempted to push him back on,
hoping that the number that appeared most often on the
display was not his full weight. It was. After a week of
wheedling, cajoling, adjusting the formula and timing of
feedings, Bean had gained just a gram.

Bean knew what he liked and didn't like, and he
didn't like his formula. Nuts and fruit were interesting
and tasty, but after a few drops of formula, he would
turn his head and push the syringe away. Knowing the
attitude of wild red squirrels, I shouldn't have been surprised; few who pass near a conifer glade can be un-
aware that red squirrels are adamant in their opinions
and not afraid to tell you about them. I have a number
of squirrel books and articles in my library, and the following quote is typical of those describing Bean's clan:
"The red squirrel is characterized by its noisy vivacity,

an impetuous inquisitiveness, and a sense of ownership that is pugnaciously maintained."

Red squirrels have reason to be feisty; to survive a northern winter they depend upon a well-provisioned larder. Like gray squirrels, they spend much of the late summer and all of the autumn in a manic gathering and storing. Conifer seeds make up the bulk of the red squirrels' winter fare. The cones must be harvested before they ripen and release their seeds. When I camped beside a fir tree one summer, I awoke each morning to the thud of falling cones. The squirrel responsible gathered this sticky treasure and stashed it in a shallow pit she had excavated and would later cover with dirt and duff. How much easier it would be to simply raid another squirrel's cache than to gather one's own! Vigilance must be maintained to thwart raids.

Since few besides red squirrels are willing or able to strip cones to eat the seeds, however, are they wasting their energy on indiscriminate rants? After all, how often do humans steal spruce cones? Cones are not the only foods stored. I first noticed red squirrels harvesting and stashing mushrooms in the dense spruce-fir forests of Nova Scotia, where I saw mushrooms in repose on many branches. Red squirrels even store and eat mushrooms that would be deadly to humans. Other foods of more general appeal are also stored; I have watched red squirrels bury nuts and even individual sunflower seeds, gray squirrel-style. They also stash fruit and nuts in bark crevices or at the base of branches. Who knows which among these delicacies might appeal to a hungry hominid?

Among my favorite anecdotes about the scrappiness of *Tamiasciurus hudsonicus,* are those recorded by Mason Walton in *A Hermit's Wild Friends* (1903). Walton's descriptions lack objectivity, but it is hard to be objective about a creature that stamps its feet and chatters when upset. In one, "Ten crows, made bold by hunger, attacked Tiny and tried to take possession of a loaf of bread. The squirrel never flinched, but stood over the bread and whenever a crow got over the deadline, filled the dooryard with feathers....The black rogues were obliged to retreat when Tiny got downright mad." In contrast to this account, Walton describes Tiny's relationship with a "towhee bunting," the type of bread thief he usually chased away. In this instance, "The bunting was eating from a loaf of bread, which was staked down in the dooryard, when Tiny appeared. The squirrel thought the bird would run away, but instead the latter set its wings and lowered its head in preparation for battle. Tiny was astonished. He sat up, folded his forepaws on his breast, and looked on the gamy little bunting with wide-eyed wonder. The bunting soon turned to the bread. Tiny brought his forepaws down hard on the ground, apparently to frighten the bird. Again the plucky little bunting set its wings and lowered its head. Again Tiny sat up and looked the fellow over. This time there was a comical expression on the face of the squirrel… That he admired the pluck of the bunting was evident by his action. He crept quietly to the opposite side of the loaf of bread and allowed the bunting to eat unmolested…" Walton wrote that Tiny not only allowed this bird to feed with him regularly, but noted that the pair greeted each other, Tiny with a chuckle, and the bunting with "something in bird language that seemed to express joy."

Stubbornness must be accepted in red squirrels, but refusal to grow? I took my problem squirrel to another rehabilitator, an authority on problem wildlife orphans of all kinds. She examined the bitsy Bean and said that he appeared to be perfectly healthy. She would try him on her formula. "All squirrels love this," she assured me. Bean tasted a drop and pushed it away. When she offered him solid food he sat up, held it in his paws and ate like a proper little squirrel. "Why, he's just a runt!" she declared, "Let's put him in with the others." We took Bean out to the large enclosure where five adolescent red squirrels were busily preparing for independence. Bean took to his new surroundings immediately, exploring the cave beneath the water bowl, wrestling with a pinecone… Soon another young red squirrel, more than twice the size of Bean, approached; the two sniffed noses, pawed at each other tentatively, and headed off in their own directions again.

Bean soon became a full-fledged, romping member

of the juvenile gang. He was released a month later, and while he was still a diminutive fellow, I have no doubt that he was fully prepared to express himself with the audacity of his tribe.

Great Expectations

~

WALTER DROVE ME TO DISTRACTION with his wanderings. Once I found him under the arm of the sofa, scrunching along in his striped pajamas. I returned him to his milkweed bouquet thinking he must be disoriented. The next evening he was gone again, and I searched the kitchen in vain. In the morning I found him hanging from his hindmost set of legs from the underside of an arched amaryllis leaf in the kitchen window. He had chosen a perfect setting to showcase his metamorphosis—still a homely caterpillar, but already vain. I kept an eye on him all day. He remained upside down with his chin tucked against his chest. I checked on him once during the night. He was still a caterpillar. In the early morning hours, in his last larval act, he gyrated until his skin split and revealed the chrysalis beneath. In the morning I found his little crisp of skin on the windowsill. The elegant pale green capsule hanging above was studded with gold—little wonder we call this butterfly the monarch.

Within the chrysalis the Walter I had fretted over had, quite literally, dissolved. The raw materials that were caterpillar would become butterfly. The new Walter would not cling tightly to the material world, but would soar aloft on fragile sails. The new Walter would not be a masticator of milkweed, but a sipper of nectar.

Walter was the great-great grandson of monarchs that spent the past winter in Mexico. In what is among the most astonishing of nature's many astonishing migrations, Walter's progenitors sailed to Mexico last fall from as far away as Nova Scotia. After spending the winter in a monarch colony in the mountains of Mexico, Walter's great-great grandmothers then managed to fly north once again. Somewhere between Texas and Pennsylvania, sometime in April, they laid their eggs on milkweed leaves and died soon after. In the course of one month, these eggs hatched into little white larvae that promptly began eating—first their egg casings, and then the leaves they hatched on. Over the course of the next two weeks they became the familiar yellow, black, and white striped caterpillars. Within a month of the eggs appearing, a next generation of monarchs took flight. Many of these butterflies continued traveling north before settling down to the business of procreation. Walter's parents, the last of the summer monarchs, were probably the fourth generation of the year.

The summer monarch generations have an easy life, but a short one. As butterflies, they enjoy drinking nectar and being gorgeous for just a month or two before their date with the reaper. Walter and the other butterflies that emerge in the fall might live for as long as eight months, but to them falls the duty of making it back to the mountains of Mexico ahead of the killing frosts, surviving the winter in the bitter cold mountains, and making the return trip north in April.

There are many mysteries and marvels surrounding the migration of monarch butterflies, and research is beginning to illuminate some of them. Consider the obvious migration question: How is it possible for such delicate creatures, easily pushed about by a breeze, to travel more than two thousand miles? Most of us learned as children just how fragile the wings of a butterfly are. Why would such wings not be in tatters part way through such a journey? Monarch researchers have found that, like the migrating hawks, these butterflies take advantage of thermals, rising with warm air columns, and then soaring on air currents at elevations as high as a kilometer above the ground. Most of their migration takes place during the warm part of sunny days. They can cover many miles effortlessly gliding high overhead.

One day, Walter's chrysalis was no longer green. I could see the folded orange wings beneath the now clear capsule. I don't know how I missed the emergence. I'd been watching all morning, but after a distraction I looked back, and there he was, his wings rumpled, clinging to the sheer remains of his chrysalis. He rocked slowly from side to side testing his new appendages. He coiled and uncoiled his long tongue and slowly pressed his wings together and then relaxed them. After about an hour of this orientation he remained still and seemed to be gathering strength for the next step. It was a couple of hours before he moved again, but when he did, it was decisively; flapping his wings, he strode up the amaryllis leaf and stepped onto my finger. I was able to admire him eye to compound eye. He was lovely! The velvety black body sported white polka dots, the brand new legs had a metallic sheen, and oh those wings! I saw no spots on the hind wings, though, and realized my mistake. Walter was female.

As I headed to the door, Walter prepared to take off. Once the sun was on her wings, she launched herself confidently skyward. She could fly! She avoided a pine tree and fluttered to land on a nearby locust. From here she will head southwest, and if she's lucky and fit, she'll soar on to Mexico. Bon voyage, Walter!

Fretful

~

"I could a tale unfold whose lightest word
Would [cause]…
Thy knotted and combined locks to part
And each particular hair to stand on end,
Like quills upon the fretful porpentine:"

—WILLIAM SHAKESPEARE, *Hamlet*

FOR THE PAST FEW WEEKS, the apple tree in my backyard has been the center of the universe for the porpentine, Fretful. I occasionally paused beneath him and made what I hoped would be perceived as friendly overtures. He would chatter his teeth fretfully and continue his activities.

Last Sunday, when I saw Fretful wandering below the tree looking for drops, I sat down nearby and rolled some apples in his direction. Fretful continued his search without raising his head, but expressed his anxiety level by raising and lowering his quills. When he wasn't fretting, his quills lay like hair on a dog. When alarmed, all quills flared. As he moved about, sets of quills would rise and fall like sports fans doing the wave.

At last he caught the scent of my apples. Quills rose and fell as he stalked, pigeon-toed, to an apple ten feet from me. He settled down, clamped the apple between his sloth-like claws and the flat pads on the bottom of his feet, and removed a few strips of peel with his incisors before launching into the flesh beneath. Porcupines peel apples! As he peeled, he flapped the spurned parts off to the side with a surprisingly long tongue. Fretful chewed with the same methodical industry as beavers, but punctuated by slurpings that made me wish I had saved some apple for myself.

Backlit and washed in a warm glow from the lowering sun, Fretful exhibited other porpentine characteristics I had not observed before. Long whiskers sprouted from above his eyes and bounced like shiny ornaments while he chewed. A halo of long guard hairs grew out beyond his quills, and, with the whiskers, must provide constant tactile information about his immediate environment.

When he finished, I expected him to wander off. To my surprise, he ambled over and sniffed my feet. He then explored the area around me. I felt each of my particular hairs stand on end as that battalion of 30,000 tiny spears passed within an inch of me. After all, if Fretful had a moment's misgiving, I would have a most unpleasant date with a pair of pliers.

~

This encounter with Fretful sent me back to my favorite porcupine book, *The North American Porcupine*, by Uldiz Roze. Roze, a biologist who has studied and admired porcupines for more than a quarter century, interweaves biological information with tales of his many interactions with these animals. On the subject of quills, for example, he describes the mechanism that allows quills to become embedded in an adversary, but not, say, the branch a porcupine leans against. A band of connective tissue, the guard spool, surrounds each quill just below the skin. When a porcupine is relaxed, this guard spool moves freely. When quills are in defense mode they are held upright by a piloerector muscle. Another set of muscles holds the guard spool in place. The pressure of impact with an adversary drives the bottom of the quills into the porcupine far enough to tear the connective tissue and the quills pull free.

Roze then shares the story of the time a porcupine managed to swat his arm. One quill had been driven in so deeply by the initial impact that it was completely buried. He observed the course the quill took as the pain moved down his arm. Two days later the quill emerged about ten inches from the initial point of entry. He was surprised that no inflammation accompanied this invasion of his person. This prompted analysis of the layer of grease that coats the quills, and he found that it contains a group of fatty acids that are known to have bactericidal properties. Roze goes on to note that one other animal is known to have a similar coating of fatty acids on its skin, a coating that is believed to deter infection. That animal is Homo sapiens.

While Fretful enjoyed my offerings, I noticed he had a distinctive odor, a sweet, piney pungency. At the time I imagined that it might be a result of the challenges porcupines must have maintaining hygiene. According to Uldis Roze, this odor is part of the porcupine early warning system. A set of short quills form a rosette around sebaceous glands above porcupine tails. When raised, these quills squeeze the glands, which secrete the aromatic compound that announces danger. These quills are barbed in a pattern that wicks the oils up and increases the surface area that broadcasts the smell.

Fretful and I had another short visit the following evening, and then for several nights I did not see him; the apple crop was finally finished. This evening, however, I spotted him again. When he noticed me, he headed up the path in my direction. I handed him an apple and he sat down beside me to eat. I did not smell his special perfume; he must have been calm. When he had eaten his fill, about an hour and three apples later, he poked around some more and then headed into the woods, leaving behind a small pile of parings and precisely excised cores. I invite you to join me at this point in the narrative:

As I lie in the grass in the growing dark typing these words on my laptop, I hear Fretful return from the woods. He walks past me and comes around to face me. I would be crazy to reach my nose toward him, I think. Surprisingly, I don't. Instead Fretful rises on his hind legs, places a forepaw on my shoulder, and reaches his face forward to mine. His whiskery nose tickles my face for several seconds, and then, satisfied that all has been said and done, Fretful returns to the woods. And what is a person to do after receiving such a blessing? I have been taken into the confidences of the porpentine Fretful and the night is exalted.

Beechcombing

⁓

THOSE EVENINGS WITH FRETFUL launched a new enthusiasm. Over a year has passed, and with Fretful as a willing guide I have learned about the lives of porcupines in all seasons. The stories will not fit in these pages, but I will include one more that took place a year later:

While porcupines have a reputation as solitary and curmudgeonly, Fretful demonstrated a keen interest in spending time with me. It may have been our differences that drew us together. Novelty has magnetism. Fretful, a heavily armored quadruped, navigated these woods and fields with sophisticated tactile and olfactory systems, while I relied on eyesight and an out-sized cranium to do the same. Still, part of what drew us to each other was recognition of kinship, for we are, all of us, made of stardust glued together by sunshine. We have risen from the same roots, governed by the same natural laws. We are, each of us, the triumphant living result of a venerable lineage of successful progenitors. This shared history enables Fretful and I to interact as familiars despite our differences. He knows that I am a companionable bene-factor and I can see that my overtures are appreciated.

The first fall of my relationship with Fretful, the beeches and oaks, the trees that determine the welfare of many creatures that face winter in these hills, produced no nuts. For bears, porcupines, gray squirrels, turkeys, blue jays, and more, the winter would be a lean time indeed. Fretful's interest in the company of humans proved a boon, for his small fan club delivered acorns from Maine and northern Vermont. Every evening he ambled to my yard where he ate as many apples and acorns as he pleased. Unlike the beavers, once he fin-ished eating he lingered, exploring the strange environs and taking an interest in my activities.

Just before Thanksgiving, he headed off to his winter quarters and his visits ceased. When the first true snow-fall came, I began to search for Fretful's winter home. At every opportunity I headed out on skis to survey porcu-pine habitat. By the time I located Fretful's winter den I had made the acquaintance of nine local porcupines, and by snowmelt was familiar with the dens and winter habits of sixteen.

Each new season required a search to locate Fretful as his diet shifted. In spring he moved to a section of forest where he fed on fungi and ash and aspen twigs. In summer he moved back to the meadows of my neigh-borhood to eat milkweed, raspberry, and violet leaves. As summer waned I hoped that Fretful would remember where he had found such a good acorn and apple crop the previous year. He did not appear in my yard, however, so I surmised that he was finding native nuts. Since oaks are more than rare in Fretful's woods, I decided he must be eating beechnuts. While this offered a starting place for my search, the woods of Marlboro are full of beech trees.

I began by surveying the beeches that I knew to be in his home range, but saw very few nuts and no sign that a porcupine had been feeding; perhaps I didn't know what to look for? I decided to visit a beech stand that is a known destination for bears and porcupines to see what evidence of porcupines eating beechnuts looked like. I could then continue my survey of Fretful's haunts with greater confidence.

I set out on one of those perfect days that October often bestows—blue and golden and mild. Zoot the goat—part cashmere, sometimes docile, and always interested in participating—invited herself along. Her soft, thick, wavy, white coat shimmered as she trotted along beside me.

Once at the top of the beech knoll I could hear someone raking leaves. I decided to do a wide circle around the sound with the hope of seeing the mystery raker. As I maneuvered among the beeches, my idea of what porcupines feeding in beeches would look like was reinforced;

beneath some of the trees, the forest floor was littered with small branches, the ends of which were adorned with the bristly husks that once enclosed beechnuts. Most of these twigs bore the mark of the porcupine— nipped off by chisel teeth. Overhead, more branches dangled in strange places, hung up in their descent.

I could not move quietly in the litter of dry leaves, but whoever was raking continued undeterred as goat and I made our noisy approach. Sure enough, at the sonic center, we located a porcupine, nose to the ground. Not wanting to startle the creature, I announced my presence from a distance of about thirty feet. The porcupine looked up and sniffed the air in my direction. He looked like Fretful, but this grove was well north of Fretful's known range. "Fretful?" I asked. The porcupine waded eagerly toward me humming an affirmative response.

Fretful embraced a proffered apple ecstatically. While he peeled and slurped and chewed, Zoot lay down in a patch of sunshine and dozed. I reclined next to the porcupine and enjoyed the perfection of the day. When Fretful had polished off three apples, I dug in my satchel for acorns. Fretful climbed up onto my lap to supervise and inspect.

Once Fretful had selected an acorn, he would support my hand from below with his paw and bite into the acorn hard enough that he could pick it up. Porcupine teeth make short work of an acorn shell. Three to five irregular shell fragments fell to my lap as each acorn was processed. The eating took more time as Fretful shaved off small pieces of acorn and chewed each thoroughly, taking a few minutes to eat an acorn.

Fretful judged about half of the acorns I offered to be palatable, and by time he completed his meal and clambered off my legs my toes had been asleep for a while. Fretful then proceeded to demonstrate how porcupines search for beechnuts on the ground— with an arcing sweep of a front leg he gently pushed aside just the top layer of freshly fallen leaves, all the while busily snuffling. When he located a beechnut, as he did on about every second or third sweep, he would pause to shell and chew. I swept away some leaves myself, and discovered that beechnuts were indeed abundant on this little knoll.

Because this area had been the locus of porcupine feeding for at least a few weeks, I felt confident that I would be able to find Fretful for as long as the beechnut crop lasted. This, however, was not to be, for there was no sign of a porcupine in that grove in the weeks that followed.

As is the way with the study of anything, as some questions are answered, new ones arise. I now know how to recognize the sign of porcupines feeding in beech stands. As for the movements of one particular porcupine during a fall season such as this, well, I have more questions. I followed the pattern of raked leaved from the beech knoll into the stream valley, but what would a porcupine find to eat under maple, ash, and birch trees? I rediscovered Fretful in a winter den in mid-December, but can't account for his activities in the intervening months. Fortunately, Fretful's den is close to my house, so I look forward to a long season of learning and of enjoying the unlikely companionship of this curious, gentle creature.

~

And so the "master story" continues. Fretful, you and I, Willow and Dewberry, Terrible Jack and Old One Eye now occupy the stage. As we play out our lives we arrange the set for those who follow. We humans bear a weighty responsibility for the form the story takes. Humanity's actions will determine which of the ancient lines of life will continue to evolve and which will disappear. I am not going to tell you how important each line might be to human economies or even to the planet's life support systems, for even though they could be, and even though some argue these are the lenses through which humans are best able to perceive the loss of nature, I believe there are things that are more elemental: things like summer evenings when the air is soft and fragrant, when fireflies flash, thrushes sing, field crickets chirp, and little brown bats swoop and flutter. I believe people would mark the absence of these things and know their lives to be poorer. We save what we love.

The Beavers of Popple's Pond was typeset in ITC Berekely Oldstyle, which was created in 1983 as a revival of Fredric Goudy's custom California Old Style™ typeface, which was originally created specifically for the University of California Press at Berkley. It is a serif font with unusually elongated descenders and a uniquely light style.

ITC Berkeley Oldstyle offers the flavor and dynamics of Goudy's original University of California Old Style without being a slavish copy. In fact, a close look reveals hints of several other Goudy designs in play: Kennerly, Goudy Oldstyle, Deepdene, and even a touch of Booklet Oldstyle.

The design is characterized by its calligraphic weight stress, smooth weight transitions, classic x-height and ample ascenders and descenders. These traits work together to create high levels of character legibility and a text color that is light and inviting.

~

PRINTED BY SPC MARCOM
SPRINGFIELD, VERMONT

TYPOGRAPHY & DESIGN
BY PATTI SMITH & DEDE CUMMINGS
BRATTLEBORO, VERMONT